The Crisis of Rural Poverty and Hunger

M. Riad El-Ghonemy argues that if current trends in market-based voluntary land reform as an alternative to government-led and market-based land reforms persist the rural poor population in developing countries will continue to rise.

Based on nearly half a century of academic and field research this valuable work presents compelling evidence on persistent rural poverty, hunger and increased inequality in developing countries over the last 30 years. The book furthers the debate with 16 detailed case studies and looks beyond the typical views of the roles of the state and the market on land reform.

The Crisis of Rural Poverty and Hunger contains comprehensive case studies including countries such as China, Korea, Egypt and Honduras and provides bases for discussions of government-mandated land reform, pro-active participation of NGOs and facilitated functions of the market mechanism. This book is essential reading for undergraduate and postgraduate students interested in the fields of rural and agricultural development, development economics and geography.

M. Riad El-Ghonemy is Senior Research Associate at the Department of International Development, University of Oxford, and Reasearch Fellow at the Department of Economics, the American University in Cairo, and Emeritus Professor, Ein-Shams University, Cairo. He is the author of several publications including *The Political Economy of Rural Poverty* (1990) and *Affluence and Poverty in the Middle East* (1998).

The Crisis of Rural Poverty and Hunger

An essay on the complementarity between market- and government-led land reform for its resolution

M. Riad El-Ghonemy

Routledge
Taylor & Francis Group

LONDON AND NEW YORK

To the memory of my father,
Riad Omar,
who, from my childhood, taught me to appreciate
rural life and to learn from the peasants

First published 2007
by Routledge
2 Park Square, Milton Park, Abingdon, Oxon, OX 14 4RN

Simultaneously published in the USA and Canada
by Routledge
29 West 35th Street, New York, NY 10001

*Routledge is an imprint of the Taylor & Francis Group,
an informa business*

© 2007 M. Riad El-Ghonemy

First issued in paperback 2013

Typeset in Times New Roman by
Prepress Projects Ltd, Perth, UK

British Library Cataloguing in Publication Data
A catalogue record for this book is available from the British Library

Library of Congress Cataloging-in-Publication Data
El Ghonemy, Mohamad Riad, 1924–
    The crisis of rural poverty and hunger : an essay on the
    complementarity between market- and government-led land reform for
    its resolution / M. Riad El-Ghonemy.
    p. cm.
    Includes bibliographical references and index.
    ISBN-13: 978-0-415-39657-8 (hardcover) 1. Land reform –
    Developing countries. 2. Rural poor – Developing countries. I. Title.
    HD1131E44 2007
    333.3′1091724–dc22
    2006032880

ISBN: 978-0-415-39657-8 (hbk)
ISBN: 978-0-415-86020-8 (pbk)

Contents

List of illustrations   xi
Preface   xiii
Abbreviations   xvii

1   The crisis: persistent rural poverty, hunger and
landlessness   1
Introduction 1
The magnitude of rural poverty 2
The dimensions of undernourishment 5
Widening gap between international commitment to reduce
     poverty and the realities 9

2   Reform of land property rights between the state and the
market: principles and concepts   16
The nature of land tenure and property rights 17
The meaning of institutional monopoly and exploitation in
     agrarian economies 19
Non-market land property transfer: inheritance rules 24
On land reform and agrarian reform 26
Neoliberalism for voluntary land reform in a private market
     economy 30
Food security and the households' power of commanding their
     food needs 33
Why focus on absolute poverty in rural development? 34
Methodological problems in measurement 37
Summary 41

3   The agricultural dimension of poverty   44
    Access to land and commanding food 44
    Key structural problems in agriculture 46
    The deficient institutional framework of agriculture 54
    Summary 61

4   Farm size and productivity: are small farms the engine for
    growth and poverty reduction?   63
    Foundation of judgement 63
    Empirical examination of efficiency in developing countries 70
    Summary 77

5   Case studies I: state-administered complete land reform   79
    Classification into complete and partial land reform 79
    Complete land reform 80
    Countries having quasi-complete land reform 112

6   Case studies II: partial land reform   118
    Some common features 118
    Egypt 119
    India 129
    Iran 131
    Morocco 133
    The Philippines 138
    Syria 140
    Summary 142

7   Market liberalization: market-based land reform   144
    Understanding the origin of the policy shift 144
    Countries' experiences 151
    *Market liberalization impacts on existing state-administered
       land reform 157*

8   Challenges and prospects   168
    *The pace of poverty reduction according to types of land
       reform 168*
    Non-land reform factors contributing to poverty reduction 171
    Suppositions challenged 174
    Prospects and challenges 178

Appendix A	183
Appendix B	186
Appendix C	188
Notes	190
Bibliography	204
Subject index	220
Name index	225

Illustrations

Figures

2.1   A conceptual illustration of the dynamics of land policy and the
      rural development process   35
5.1   Lorenz curves for changes in the degree of inequality in land
      distribution in South Korea, 1930, 1945, 1970 and 1980   97
5.2   Lorenz curves for changes in the size distribution of landholdings
      in Iraq, 1958, 1971 and 1982   109

Tables

1.1   Estimates of the rural poor and agricultural landless wage workers
      in 14 selected developing countries, 1979–90   6
1.2   Correlation between the percentage of undernourished and other
      characteristics of poverty   10
1.3   Regional change in the World Bank allocation for agriculture and
      rural development, 1977–87   11
2.1   A hypothetical index of land and income distribution for countries
      X and Y   36
2.2   Hypothetical example of the consumption behaviour of the poor   41
3.1   Performance of agriculture in developing countries, 1965–2003   47
3.2   Food production in eight developing countries with egalitarian
      agrarian systems, 1960–2003   49
3.3   Changes in arable land and pressure of agricultural workforce on
      land in 13 developing countries, 1970–2002, and projection for 2010   53
3.4   Changes in the concentration of size distribution of landholding in
      23 developing countries, 1950–2004   57
3.5   Inequality of land distribution and agricultural GDP growth rates,
      1960–2000   60
4.1   Mozambique state farms' marketed crops, 1981–83   70
4.2   Rural development indicators in Kenya, 1979–2003   76

5.1   Estimated redistributive scope of state-administrated land reform in
      22 developing countries (excluding settlement schemes), 1915–94   83
5.2   Gini coefficient of pre-reform land concentration in four Chinese
      regions, 1929–36   84
5.3   Selected indicators of agrarian and human capability changes in
      China, 1930–85   86
5.4   Distribution of land ownership – all Korea, 1927 and 1930   92
5.5   Stability in the size distribution of landholdings after land reform in
      South Korea, 1960–80   95
5.6   Different findings for annual rates of growth for the value of
      agricultural output in South Korea   96
5.7   Available estimates of rural poverty in South Korea   99
5.8   Changes in the distribution of landholding in Iraq, 1952–82   105
5.9   Changes in agricultural income and rural quality of life in Iraq,
      1948–80   107
5.10 Post-reform index of food production in Mexico, 1956–68   113
6.1   Areas acquired for redistribution by Egyptian land reform laws,
      1952–69   121
6.2   Changes in the size distribution of landownership in Egypt,
      1951–84   122
6.3   Average yields per feddan in three land reform districts and
      national averages, Egypt, 1954–64   125
6.4   Changes in household asset ownership and per capita income in a
      land reform area, Gabaris, Egypt, 1953–73   126
6.5   Average yield of main crops in three reformed areas of Egypt
      compared with the national average   126
6.6   Estimates of inequality (Gini coefficient) in income/expenditure
      distribution in rural areas of Egypt, 1940–82   127
6.7   Prevalence of poverty in rural Egypt, 1949/50–82   128
6.8   Variation in poverty reduction between all India and Kerala,
      1956–78   130
6.9   Comparison of quality and quantity of life between Kerala and
      all India, 1970–81   131
6.10 Post-land reform distribution of landholdings by size in Morocco,
      1974   137
6.11 Size distribution of agricultural landholdings, 1970 and 1994,
      in Syria   141
7.1   Per capita income of rural and urban households, 1957–90   159
7.2   Landholding distribution in Egypt, 1981–2000   160
7.3   Average land values and daily wages in Egyptian agriculture,
      1930–98   162

Preface

When I was writing my PhD thesis on 'Land Reform and Economic Development in Egypt' at the University of North Carolina, Raleigh, USA, it was unthinkable that, half a century later, I would be arguing that the poverty-alleviating state-mandated redistributive land reform should be not replaced but complemented by the recently promoted market-based voluntary land reform. Rather, both types of land policy should supplement each other, owing to the increasing magnitude of rural poverty, malnourishment, landlessness and inequality of land distribution. Their complementarity for bringing about social justice is needed because of the interdependence between an effectively functioning market and governmental actions. It is this emerging obsession with a universally prescribed single path of market-based land policy – to the exclusion of government intervention and irrespective of country-specific variations in initial agrarian systems and cultural values – that has triggered the writing of this book.

The impulse to write has been sharpened by four consecutive events. The first was the commissioned study on the World Bank's induced and heavily subsidized market-assisted land reform schemes in a few countries for the UN Research Institute on Social Development (UNRISD), Geneva, 1999. The second was the kind invitation from V.K. Ramachandran, Social Sciences Division, Indian Statistical Institute, Kolkata, to present a paper on the land market approach to poverty reduction at a conference held in January 2002. The third was a series of discussions with Alan Jarvis, publisher at Routledge, on our shared concerns about worsening land concentration, rural poverty, undernourishment and increasing numbers of the unfairly disadvantaged wage-dependent landless workers in most developing countries, amid waning international commitment for genuine land reform. Our discussion also included the current acrimonious debate on pro-market and anti-government intervention to rapidly reduce inequality and poverty in rural areas of developing countries. Last was the Food and Agriculture Organization's (FAO) invitation in September 2005 in Rome, to write a paper on the rural development challenges arising from the recently published data on increased numbers of undernourished rural people and landless poor while the agricultural labour force is rising fast and the aggregate supply of cultivable land is steadily diminishing, resulting in their harmful effect on productivity and social justice in most developing countries. Concurrently, in the post-economic reforms and market liberalization of the 1980s and 1990s, a body of accumulated evidence on slow agricultural and total economic growth has pointed to deep-seated increased landholding inequality as the main attribute.

I should not be misunderstood in what I am trying to say. This book is not a critique of the market-based, voluntary, landed property rights transfer because it has worthy goals, but it is against its promoters' exclusion of the over six decades of proven poverty and hunger-reducing redistributive land reform as a direct approach to anti-poverty rural development. I believe in this interdependence not in the abstract but as manifested by several countries' success and as a result of my own 52 years' experience at village, national and international levels, including 26 years with the FAO of the UN. I am reporting in this book on reliable findings of carefully conducted case studies and cross-country analysis as I am convinced of their contribution to the current debate.

I write this book as a development economist. Being a trained agricultural economist, I have been concerned with the institution of land tenure arrangements and related command over household food needs in my home country, Egypt, and many other developing countries in which I have worked and carried out field studies, particularly in Latin America and the Middle East. Through research and practical work, I have realized that poverty and social injustice in agriculture cannot be separated from the structural forces operating in the national and global economy. Conceptually, this perceptive expansion became clearer during my seminars in the 1980s and 1990s to which I was invited by the Universities of Cornell, Cairo and Ein-Shams (Egypt), North Carolina, Arizona (USA), Glasgow and at several colleges of Oxford University. Also, the interdependence of agrarian and national structural adjustment has become clearer during my academic affiliation and research work at the Institute of Development Studies, University of Sussex, in 1986–87 at the kind invitation of the late Sir Hans Singer and, since 1987, at the Department of International Development, at Queen Elizabeth House, University of Oxford, at the kind invitation of the late Professor George Peters.

The book is structured as follows. Chapter 1 introduces the magnitude of the problem addressed throughout the volume and the relevant trends observed since 1950. It presents the accumulated evidence about the present crisis of rural poverty and hunger manifested in the increasing prevalence of rural poverty, chronic undernourishment or hunger, and landlessness, and it examines their development linkages and characteristics. The chapter also shows how, in reality, the contraction of international aid to developing countries' agriculture and rural development contradicts the declared commitments. Chapter 2 defines the terms used, examines the principles behind the roles of the state and the market, with special emphasis on social justice and neoliberalism, and sets the analytical frame of discussion. Chapters 3 and 4 investigate the agricultural dimension of rural poverty, both physically (food production) and institutionally (land tenure, farm size and inequality in the size distribution of landholding/ownership), and ask the questions explored in the rest of the book. Empirical evidence on state-administered redistributive land reform presented in case studies and derived from my own field work are the subject of Chapters 5 and 6. Chapter 7 explores the foundation of the shift away from redistributive land reform towards market-orientated land policy, and reviews the experience gained from pilot projects in a few countries. It also examines the privatization of centuries-long customary or communal land in sub-Saharan Africa and in some Latin American and Asian countries. The last chapter presents the prospects and challenges ahead. It attempts to answer the questions addressed in the preceding chapters, and argues for the complementarity of both methods of securing land access in the light of the long experience gained and the distressing magnitude of poverty and social injustice in the rural areas of most less developed countries (LDCs).

I am grateful to Barbara Harriss-White, who read an early draft of the concluding chapter and offered valuable comments. I offer special gratitude to the FAO of the UN, in which I served 26 years and learned a good deal from my colleagues and from the farmers in the many countries that I visited, carried out studies and worked. My thanks are recorded to the UN Institute for Research in Social Development (UNRISD, Geneva) for their permission to use some data contained in my study *The Political Economy of Market-based Land Reform*, 1999. I acknowledge with thanks the provision of useful information on the 2006 Bolivian land reform policy by the Embassy of the Republic of Bolivia in London.

This work has not been supported by grants from any source. But, as mentioned earlier, it has been encouraged and actually proposed by Alan Jarvis, publisher at Routledge within the Taylor & Francis group. The progress of its publication has been indebted to the keen interest and patience of both Amber Buckley, assistant editor, and Andrew R. Davidson, production editor. The complex subject of this book has required much searching for material in libraries. I am most grateful to the generous assistance of several librarians at the American University in Cairo, Queen Elizabeth House library, before its closure in September 2005, helpful staff at the Political Science and Philosophy Reading Room (PPE) of the Oxford University Bodleian Library, the Middle East Centre of St Antony's College and the Social Science Library at the Department of Economics, University of Oxford. I would also like to thank Julia Knight, Denise Watt, Penny Rogers and Marina Kujic for their kind administrative support at Queen Elizabeth House, Oxford. My deep thanks go to Jane Gaul for her efficiency in word-processing most of the manuscript and the entire set of tables. Without her efficiency, I would not have been able to complete this manuscript in time. Also, many thanks to Mohammad Shihata, who processed three chapters in Cairo. Lastly, my wife, Marianne, has patiently read some of the barely legible drafts. She and my daughter Samira have put up with my long absences abroad and solitary work at home.

Riad El-Ghonemy
Queen Elizabeth House
University of Oxford
Mansfield Road, Oxford, UK

Abbreviations

AID	Agency for International Development of the US Government, Washington, DC
BMR	basal metabolic rate
CIDA	Inter-American Committee for Agricultural Development, Washington, DC
CLR	complete land reform
CPE	centrally planned economies
CPI	consumers' price index
DAC	Development Assistance Committee of the OED
ECOSOC	United Nations Economic and Social Council, New York
ESCWA	UN Economic and Social Commission for West Asia, Beirut
FAO	Food and Agriculture Organization of the United Nations, Rome, Italy
GDP	gross domestic product
GFCF	gross fixed capital formation
GNP	gross national product
ha	hectare = 2.4 acres
HYV	high-yielding variety
IFAD	International Fund for Agricultural Development, Rome, Italy
IFPRI	International Food Policy Research Institute, Washington, DC
ILO	International Labour Organization, Geneva
IMF	International Monetary Fund, Washington, DC
LC	land concentration
LDCs	less developed countries or the World Bank's classified developing countries, excluding the members of the former Soviet Union
MDG	millennium development goal of halving poverty and hunger by 2015
MLR	market-based land reform
MNC	multinational corporation
NGOs	non-governmental organizations
ODI	Overseas Development Institute, London

OECD	Organization for Economic Cooperation and Development
PLR	partial land reform
SALR	state-administered land reform
UNDP	United Nations Development Programme, New York
UNESCO	United Nations Educational, Scientific and Cultural Organization, Paris
UNICEF	United Nations International Children's Fund, New York
UNU/WIDER	United Nations University/World Institute for Development Economics Research, Helsinki
WCARRD	World Conference on Agrarian Reform and Rural Development, Rome, 1979
WHO	World Health Organization, Geneva

1   The crisis

Persistent rural poverty, hunger and landlessness

Introduction

In this first decade of the twenty-first century, one of the world's major development problems is the present crisis of rural poverty and, related to this, chronic malnourishment or hunger and increasing landlessness. They remain staggeringly high on a persistent basis in most developing or less developed countries (LDCs).[1] Their consequential threats of ill-health and social unrest and productivity loss are real, requiring urgent joint actions by the state and non-governmental social groups (NGOs) and facilitated market mechanisms. The dictionary meaning of crisis as 'time of great difficulty', 'marked change in symptoms' and 'defining moment' is relevant to the magnitude of the present distressing state, the characterization of which and definition of their linkages is the task of this chapter.

In order to appreciate the potential benefits of resolving the crisis, we need to note that the rural population in many LDCs is still the majority, and agriculture, the provider of food and the bulk of rural jobs, continues to be a key sector in their economies; agricultural land plays an important part in determining political advantages and social status. In addition, inequality of landholding distribution has worsened, and educational opportunities are limited, particularly for rural women. Coupled with rationed agricultural credit and scarce irrigation water, these are important assets whose accessibility to the poor is constrained and almost absent in many cases, thus inhibiting both economic growth and poverty reduction.

It is hardly surprising that there is a growing concern over the socio-political implications of these serious development problems, interacting within a crippled agriculture. Since the early 1990s and following the wave of structural adjustment and price liberalization programmes, poverty reduction has received renewed importance in the international development agenda. Examples include the World Bank's *World Development Report 2000/01*, the *Rural Poverty Report 2001* of the International Fund for Agricultural Development (IFAD), the United Nations Millennium Declaration (2000) on halving poverty and hunger by 2015 (see Appendix B) and the FAO's *State of Food Insecurity in the World 2004*.

Amid this growing concern, the policy debate on how best poverty in rural areas can be tackled is becoming bitter and sometimes even full of hatred between the promoters of pro-market voluntary land purchase and opponents of this path, who argue for an active government interventionist policy to break up the continued land concentration, which is the basic determinant of landlessness and poverty and, in turn, undernourishment and ill-health. The present author believes in both sides of the debate not as alternatives, but as complementary. I also adhere to the idea that the role of agricultural growth, investment in human capabilities and popular participation is central, and that this trinity should be integrally reinforcing each other in a poverty-reducing rural development strategy. However, within a largely political scope, this debate has compounded and narrowed the multidimensional nature of the rural poverty crisis, and it tends to distract attention from the central issue of increasing inequality in the distribution of agricultural land that slows down economic growth and deprives poor people of command over their food needs and labour employment.

The chapter begins with an attempt to understand the magnitude and meaning of falling international aid to agriculture and rural development, leaving the investigation of land distribution inequality and its impact on agricultural output growth and food productivity to Chapter 3, and the relationship between farm size and productivity of both labour and land to Chapter 4. Understanding the effects of the scale of land redistribution on the pace of poverty reduction is the purpose of Chapters 5, 6 and 7, presenting 16 case studies supported by the results of my analysis of data from 21 countries (El-Ghonemy 1990a, 1993).

The magnitude of rural poverty

It is easy to talk about rural poverty in the abstract as an evil which should be eliminated. Despite the progress made in poverty measurement, and in spite of the remarkable progress made during the last two decades, it is not so simple to measure the actual numbers of poor people. To move from the abstract to the concrete, it is necessary to identify the poor and understand the extent of rural poverty in the developing countries, so defined. The rural poor form an economically and socially heterogeneous group made up of hard-working and risk-taking tenants, sharecroppers, small landholders, especially in rainfed agriculture, wage-dependent landless labourers, pastoralists and artisan fishermen. Within each group, there are the very poor or destitute, who require special development programming. There is also an important difference between persistent and seasonal poverty. The latter is due to drought or floods or earthquakes, and may persist if not speedily tackled. However, in this study, we focus on absolute poverty, meaning absolute deprivation of dignity and certain basic necessities of life, the most obvious being a daily stipulated minimum amount of calories per person required for a minimum healthy life according to age, occupation and sex.

Several countries have established national, rural and urban poverty lines to determine the number of households or individuals who are poor, i.e. those who, in physical or income terms, live below the established standard. In estimating a poverty line, some allowance is made for spending on non-food items, such as clothing, housing and fuel, as well as on education and health. Clearly, nutrition is the most essential of human necessities for life. Therefore, the poverty line consists of two components: a food- or nutrition-based line corresponding to the cost of the minimum food requirements below which the person is called by some researchers a destitute or ultrapoor; and the cost of minimum non-food requirements. The aggregation of those falling beneath the poverty line in different years conceals: (i) the occupational categories of the rural poor; and (ii) how poor are the poor, meaning how much the average income/expenditure of the individual rural poor falls short of the poverty line known as the poverty gap, which measures the depth of poverty. For cross-country comparison, we need to use the same head count measure, which is the number of the poor as a percentage of the total rural population. Later, the World Bank's internationally simplified poverty line of 1 or 2 dollars per person per day will be discussed.

How many are the rural poor?

In El-Ghonemy (1986: Table 1.1), I compiled available estimates of rural poverty (head count) from 60 developing countries for the period 1975–82, with a varying basis of calculation (income/expenditure), rural/urban definition and prices for costing the components of the poverty line, i.e. for translating such physical constituents as food into monetary value. Given these limitations, the estimated total number was 671 million out of the total rural population of 1,321 million or 45.4 per cent of the total in the 60 developing countries. This attempt has continued in 1990 by refining and updating some estimates (El-Ghonemy 1990a: Annex Table A) and adding four other developing countries making a total 64 countries representing 90 per cent of the total population of all LDCs. Accordingly, the number of rural poor was estimated at 767 million persons around 1985. For the reader's easy reference, this table is reproduced in Appendix A at the end of this volume.

Here again, aggregation hides away important variations. Of the 64 countries, only 14 had a 'low' prevalence of rural poverty (below 30 per cent).[2] At the other extreme, 34 countries suffered 'high' poverty prevalence of 50 per cent and over. Classification by region shows that 16 are in Africa, 10 in Latin America, four in Asia and three in the Middle East. Of the 767 million rural poor, nearly 70 per cent are concentrated in seven countries, mostly Asian: India (266 million), China (60 million), Bangladesh (56 million), Indonesia (52 million), Nigeria (40 million), Brazil (26 million) and Pakistan (24 million).

Calculating the rural poor has continued since then. In 1992, IFAD published data on poverty estimates (1988–90) in 114 countries, including my prepared work on North Africa and the Middle East.[3] The total number of rural poor was 939,481 million persons, constituting 36 per cent of the total rural population and 80 per cent of the total number of poor people (rural and urban) in these 114 countries.[4] Geographically, they are distributed as follows: 663 million in Asia, 204 million in sub-Saharan Africa, 76 million in Latin America and the Caribbean, and 27 million in North Africa and the Middle East. The highest proportion of the total rural population living in absolute poverty is 60 per cent in sub-Saharan

Africa and the lowest is in the Middle East, 26 per cent, followed by Asia, 31 per cent (IFAD 1992: Appendix Tables 2 and 6). Examining changes over the period 1965–90 in 41 developing countries with comparable data, the evidence indicates that, although 11 have made progress in alleviating both the percentage and the numbers of rural poor, 30 countries have failed, resulting in a substantial increase in the absolute number of rural poor in the sample of 41 countries from 511 million in 1965 to 712 million in 1988 (IFAD 1992: Table 3.11).

In the meantime, since its 1990 *World Development Report*, the World Bank has contributed to international comparison of poverty. First, in cooperation with the UN, it developed the ICP/PPP (International Comparison Program and Purchasing Power Parity) by computing total gross domestic product (GDP) and per capita income in terms of equivalent dollar prices in every country. Second, using a USA dollar-based poverty line, e.g. 1 or 2 dollars a day per person, the IFAD (2001) estimates the total number of rural poor in developing countries at 900 million persons. Thus, in this rapid globalization, counting the poor is internationally dollarized. In this IFAD study, the agricultural landless wage workers are identified everywhere in developing countries. In Africa, however, small farmers in rainfed areas and nomadic pastoralists are more dominant than the landless who began to grow in numbers after the introduction of land property privatization (see Chapter 7).

Pastoralists and nomads

Estimates of the number of pastoralists and nomads in developing countries, like those of the number of landless in agriculture, differ according to the definitions used. First, a distinction may be made between the terms 'pastoralism' and 'nomadism'. Although both pastoralists and nomads depend for their livelihoods mainly on livestock maintenance and natural forage, nomads require mobility because of seasonal rainfall-based change. To simplify the discussion, both are treated here as one group.

The world's pastoralists and nomads live mainly in Africa and the Middle East. In El-Ghonemy (1986: 18), they were estimated at nearly 15 million in Africa south of the Sahara, 11 million in the Middle East and between 3 and 5 million in China, India and Mongolia, totalling about 29–31 million. They still represent a high proportion of the rural population of countries such as Somalia (65 per cent), Mauritania (80 per cent), Mongolia (60 per cent) and the Sudan (20 per cent). They are vulnerable to adverse environmental conditions, which can cause a high death rate among stock, thus reducing their asset base. This vulnerability has been tragically evidenced during recent droughts in West Africa and Sudan. In both cases, many pastoralists and nomads who had lost or sold off their animals later died from famine. Because of their mobility, pastoralists and nomads have not received or have limited access to education, primary health care and social amenities. Of course, not all pastoralists or nomads are poor. In a study of Somalian nomads, it was found that they were no worse off than small farmers (holding on average 0.15 hectares) in terms of income, but had larger family size, less access to education and agricultural services. Forty-one per cent were below the nutrition-based poverty line. The largest families of nine adult equivalents were all poor, having 14 cattle and sheep each, compared with 49 among the rest (Tyler 1983: 36).

Landlessness: who are the landless and how many are there?

From agricultural landholding data in 90 developing countries, the FAO study of 1980, '*Agriculture Towards 2000*' (p. 88) estimated the number of mini-land-holders (holding a small fraction of a cropped one acre or one hectare) and land-less households (not owning and not renting arable land and wage dependent) at 167 million; this figure was projected to be about 220 million in the year 2000. In El-Ghonemy (1990a: 20), wage-based landless agricultural households were estimated at 180 million in 1985. By 1992, the IFAD study (Table 3.18 and Appendix Table 6) estimated the landless at 24 per cent of the total rural population or 226 million, reaching 39–44 per cent in Latin America and as low as 4 per cent in South Korea. The study states, 'the situation seems to be deteriorating in many countries in terms of the size and proportion of the population which are landless' (p. 47; see also Sinha 1984).

Recognizing both the lack of a uniform definition of landlessness and the land-less workers' possible earnings from non-farming sources, I have compiled data on landlessness and rural poverty proportions in relation to the total rural popula-tion in the late 1970s and during 1980–90 for 14 developing countries, and these are presented in Table 1.1. The purpose is to understand the possible order of magnitude and changes in the two variables, grouped into high poverty (40 per cent and over) and low poverty (below 40 per cent). It is apparent that both vari-ables move together in most cases, i.e. without statistical measurement, they are positively correlated. However, it is alarming indeed that, in eight of the 14 coun-tries (57.2 per cent), a high level of landlessness has persisted: landless workers represent over one-fifth of the total rural/agricultural population, reaching 34 per cent in the Philippines and 39 per cent in Brazil.[5] Because of its significance, their relationship with the inequality of the landholding distribution is fully examined in Chapters 3, 5, 6 and 7 and in the concluding Chapter 8.

We have detailed the discussion on agricultural landless workers for the pur-pose of indicating the importance of understanding the occupational categories of the rural poor prior to policy design with regard to land access, which is the focus of this study. Moreover, if we trace this category in household nutrition and education surveys, the landless are the most likely to suffer from undernutrition, illiteracy and absolute poverty, particularly among adult females and large-sized households with many dependent children and those headed by women.

The dimensions of undernourishment

It is clear from the country-specific data on the prevalence of rural poverty exam-ined by El-Ghonemy (1986: Ch. 2, 1990a: Ch. 1), as well as those prepared for the

Table 1.1 Estimates of the rural poor and agricultural landless wage workers in 14 selected developing countries, 1979–90

Countries in descending order of poverty	Estimated percentage of rural population in absolute poverty and year of estimate (1)	Estimated landless households as a percentage of the total agricultural population (2)	Poverty and landless workers as a percentage of the rural population 1988–90	
			Poverty (3)	Landlessness
High poverty (over 40%)				
Bangladesh	78 (1982)	31	86	20
Brazil	67 (1980)	39	75	39
Honduras	58 (1980)	33	55	26
Venezuela	56 (1980)	27	58	27
India	51 (1979)	30	42	30
Kenya	45 (1979)	15	55	13
Indonesia	44 (1980)	36 (Java)	27	15
Philippines	42 (1982)	37	34	34
Lower poverty (less than 40%)				
Pakistan	39 (1980)	31	29	30
Thailand	34 (1979)	10	34	15
Egypt	28 (1979)	24	25	25
Sri Lanka	26 (1981)	19	46	22
Jordan	17 (1979)	7	17	3
South Korea	10 (1980)	4	11	4

Sources: Columns 1 and 2, El-Ghonemy (1990a: Table 5.4). Column 3, IFAD (1992: Appendix Tables 2 and 6).

1996 World Food Summit and the IFAD 2001 *Rural Poverty Report*, that most of the undernourished live in rural areas and that undernourished people are almost always poor, simply because food expenditure represents about 70–75 per cent of the household income or expenditure. Likewise, in the anthropometric surveys, higher mortality rates are recorded among rural children than among those living in urban centres.

Thanks to the collective efforts of nutritionists, economists, statisticians and the medical professionals, FAO and WHO have been able to measure the realities of individuals' nutritional status, particularly of the rural population in the developing world. Examples are the FAO periodical *World Food Survey*, an FAO and WHO jointly organized International Conference on Nutrition, 1992, and the World Food Summit, 1996. For example, in its *Fifth World Food Survey* published in 1985, FAO fixed *two* per capita requirements levels, 1.4 and 1.2 basal metabolic rate (BMR), taking into account variations between individuals. Using the lower cut-off point, the results show that 237 million people in the 33 low-income countries and 335 million people in the total of 98 developing countries were surviving at the absolute minimum requirement of 1.2 BMR.[6] Using the upper limit of requirement, we find that almost 500 million people were suffering from undernutrition (see also FAO 1996b).

These results can only suggest the enormity of undernutrition in developing countries, particularly among their rural populations. The 1992 International Conference on Nutrition recognized that, although average daily calorie supply per person is a useful indicator and both average and total calorie supply have increased since 1971, these measures do not indicate *actual* amounts of food consumed or food intake. Therefore, country, regional and international food supply estimates should be much higher because of distributional inequality and the losses and wastages (averaging 40 per cent) occurring between production at the farm level and consumption at the household level as well as those caused by the ongoing civil unrest and frequent droughts. Besides, the Conference pointed out that only about 40 per cent of total cereal production is consumed by people and the rest is consumed by livestock and reserved as seeds.

Accordingly, it was estimated in 1992 that, in all developing countries, 786 million people were chronically undernourished or hungry (they regularly failed to have access to enough food to meet their minimum dietary needs for an active healthy life). The estimate made by FAO (2001) for the year 1999 was 777 million, which represented a slight reduction of only 9 million or 1.1 per cent over a period of 7 years. Later in this chapter, I shall return to this point when I examine the prospects for meeting the established world development goals of halving undernutrition and poverty by the year 2015. In addition to these hundreds of millions in 1992 and 1999, there were tens of millions of people being seasonally undernourished. On the whole, the average proportion of the undernourished to the total population of developing countries estimated by the 1992 International Conference on Nutrition was 20 per cent, but much higher at 33 per cent in Africa (FAO, WHO 1992: Table 2). At this juncture, we should note that the World Bank's estimate of all poor people in developing countries in 1990 was 1.27 billion persons and, for the same year, IFAD estimated the number of rural poor at 939,481 persons. This estimate of the undernourished is a much higher proportion of the total poor than the share of rural people in the total population (nearly 39 per cent in developing countries in 1990), suggesting the concentration of the poor in rural areas.

The 1996 World Food Summit emphasized that 841 million people were suffering from undernutrition, owing primarily to a widespread economic inequality of their access to food that 'explains the persistence of the high number of undernourished people, in spite of the production improvements' (FAO 1996a: para. 2.41). It stressed further that most of the undernourished are in countries depending on rice, cassava and yam, mostly in Asia with the exception of China. On food production, the Summit stressed the fact that increased agricultural and food production per head of the agricultural workforce and total rural population in developing countries is essential for poverty reduction and that 'the main cause of chronic undernourishment or hunger has been the inability to reduce poverty in these countries' (FAO 1996a: para 2.62).

The health dimension

Ill-health of the rural poor is a key manifestation of poverty and undernutrition. It reduces the individual's productivity and robs the economy of the poor people's unrealized potential. Importantly, it prevents the children of poor rural households receiving the full benefit of education. Thus, undernutrition is more than a simple quantitative measurement problem of amounts of food supply. Measurements of average per head and total calorie supply may unwittingly conceal the health dimensions of hunger: physical growth, low birthweight, vulnerability to illness (tuberculosis, pneumonia, meningitis and shorter life expectancy). In El-Ghonemy (1986: 23), evidence from WHO country field studies shows that 54 per cent of rural Asian and 26 per cent of rural African children under 5 years old were undernourished according to anthropometric surveys (actual weights and heights of individual children compared with standard growth patterns), reaching 75 per cent in rural Bangladesh, with a much higher mortality rate than in normal children.

The situation is still frightening. We noted earlier that the estimated prevalence of the total undernourished population in developing countries was 20 per cent in 1990 and improved to 17 per cent in 2002. According to FAO (2004b: Table 2) undernutrition in children under 5 years old in 2000 was between 25 and 48 per cent in developing countries whose average prevalence of undernutrition was 5–19 per cent, and child mortality was very high at between 130 and 207 per 1,000 live births in eight developing countries (Benin, Burkina Faso, Ivory Coast, Mauritania, Myanmar, Nigeria, Swaziland and Uganda). The number rises to 28 countries with very high child mortality in the category of undernourishment prevalence of 20 per cent and over. The same study tells us that the total number of people undernourished in developing countries was 814.6 million in 2002. Compared with 797 million in 1995–97, this represents a deterioration in nearly 18 million persons' nutritional condition in only 5 years.

Labour productivity/income losses from undernourishment

Undernourishment not only causes ill-health leading to premature death, it also reduces adult physical and mental productivity and earnings during a lifetime. In addition, it decreases national income as a result of premature death of the thousands or millions of undernourished people. Empirical studies and economic loss estimates are now available on these relationships. The estimates vary according to the valuation coefficient used for an individual loss and for the social cost. By social cost is meant the sum of money paid to restore health and to compensate for loss of productivity as well as lives and earnings opportunities, resulting from undernutrition and related premature death. Despite their limitations, these estimates are useful as an expression of an approximate indication of the order of magnitude of individual loss and social cost to the nation.

A pioneering effort was made by Murray and Lopez (1995). In their global study for the World Health Organization (WHO), they established a methodology known as DALY (years lived with disability) for estimating the earning loss from the reduced number of years of life (shorter life expectancy) due to premature death and number of years lived with disability caused by ill-health. The costs of dealing with nutrition-based illness and disability caused by undernutrition include: the direct medical costs for the treatment of anaemic underweight pregnant women and deliveries of underweight babies; treatment of children's physical illness (e.g. diarrhoea and anaemia) and mental retardation. On the other hand, indirect costs include low productivity and wage payments; missed schooling years and reduced learning capacity of children; as well as their losses in height and weight that lead to loss of lifetime earnings. UNICEF (1998) estimated that 55 per cent of the nearly 12 million deaths each year of undernourished children under 5 years old are in rural areas of developing countries. Among the recent studies relevant to our study is the work of Horton [1999 (in FAO 2003a)] and Arcand (2001). The former, for instance, carried out empirical studies in four Asian countries (Bangladesh, India, Pakistan and Vietnam) and from his findings made an estimate of national economic costs at between 2 and 4 per cent of their national income (GDP).[7]

Undernutrition and its multidimensional correlation

So far, I have discussed each of nutrition, health, education and the scale of poverty prevalences separately. But, in the lives of the rural poor, nutrition, health and education status are intimately connectedand affect their well-being simultaneously. The multidimensional nature of poverty characteristics has been confirmed by a cross-country analysis based on FAO estimates of the percentage of undernourished and on other socio-economic indicators in 95 countries. The results presented in Table 1.2 show significant association between the extent of undernutrition, low per capita income or expenditure, access to safe drinking water, infant mortality and adult illiteracy. The health dimension represented by the two variables, life expectancy and access to safe drinking water, is negatively and strongly correlated, i.e. the lower the undernourished person's level of hunger, the higher is life expectancy. Thus, halving the number of the rural poor and the undernourished by 2015 is not going to be achieved through income growth alone, as claimed by the neoclassical economists.

Widening gap between international commitment to reduce poverty and the realities

This is the last component of understanding the prevalence of poverty, undernourishment and landlessness in rural areas. In my judgement, there has been hypocrisy in dealing with poverty and undernutrition reduction. Since the declaration in 1979 by governments of all developed and developing countries meeting in Rome at the World Conference on Agrarian Reform and Rural Poverty (WCARRD) to 'quickly eliminate undernutrition before the year 2000, and to speedily realize equitable distribution of land', there have been significant commitments proclaimed by world leaders at several summits during the 1990s to halve poverty

Table 1.2 Correlation between the percentage of undernourished and other characteristics of poverty

Other characteristics of poverty	Correlation coefficient
Adult illiteracy rate	0.5
Infant mortality rate	0.5
Life expectancy at birth	−0.5
Percentage of households with access to safe water supply	−0.6
Productivity/income (per caput)	0.4

Source: FAO Statistics Division, Rome, 1984, in El-Ghonemy (1986: Table 2.7).

and undernourishment by the year 2015. In the meantime, the rich, powerful 21 member countries of the Development Assistance Committee (DAC) of the Organization for Economic Cooperation and Development (OECD) committed themselves at the World Summit for Social Development in Copenhagen in 1996 to provide official aid to developing countries at 0.7 per cent of the members' national income (GNP), and they adopted the same goal of halving poverty and hunger by 2015. But alas, 1 year later in 1997, total official development assistance declined by 14 per cent (World Bank *Development Report* 2001: 190). These scattered and sometimes confusing international commitments were consolidated in the millennium development goals (MDGs) and were adopted by all heads of states attending the UN General Assembly in September 2000. Yet, by 2006, when the numbers of the rural poor and the undernourished had increased, total official development assistance had declined sharply in real terms to 0.2–0.3 per cent of the donors' national income, against their own repeated commitment of 0.7 per cent. This striking fall occurred despite the sustained affluence of the DAC 21 member countries and in spite of their agreed focus of development aid on the poor. It is an indication of the lack of political will.[8]

What is worse is the drastic decline in the share of agriculture and rural development, where poverty is embedded, from 20.2 per cent in 1987–89 to 12.5 per cent in 1998 (OECD/DAC *International Statistics*, 2000). Most alarming in this regressive tendency is the change in the World Bank's rural poverty-relating lending and development aid since the 1980s. It is alarming because, since the early 1970s, the World Bank has been the leading international institution for the study and tackling of world poverty. In Chapters 2 and 7, I shall attempt to explain the shift with regard to state-led redistributive land reform but, in this section, the change in aid to rural/agricultural development is examined. In 1986 and 1987, the share of the 'agriculture and rural development' sector in the World Bank's total loans and field operations was far below its annual average in 1977–81, particularly in Asia and Africa, where rural poverty is concentrated. Apart from this regional decline, lending for rural development (poverty-orientated projects) as a percentage of the total agricultural sector fell from its 1977–79 level of 52 per cent to 29 per cent in 1983–85. The regional change, based on data calculated from the World Bank 1986 and 1987 *Annual Reports*, is shown in Table 1.3. I find it difficult to update these comparable data for two reasons: first, the separation of

Table 1.3 Regional change in the World Bank allocation for agriculture and rural development, 1977–87

Annual average allocation by region	1977–81	1986	1987	Percentage change in 1987 over 1977–81
East Asia and Pacific	33	15	13	−20
South/Eastern Asia	40	36	10	−30
Southern Africa	32	20	28	−4
Western Africa	38	21	23	−15
Latin America	23	41	23	0

Source: The World Bank Annual Report (1986 and 1987).

lending for agriculture as a sector from rural development, being a theme; second, there are changes in regional coverage (see, for example, the World Bank *Annual Report* 2002, Ch. 2, Table 2.2 of Vol. 1).

Why were concerns for agricultural/rural development and poverty-reducing, 'genuine' land reform and greater equality in land distribution short lived? The answer is to be found in the *World Development Reports 1982–7* and in the heading of an article appearing in the World Bank's *Research News*, 1985. It reads, 'The World has changed, so has the Bank'. But to what fundamental changes since the mid-1980s does the Bank refer? Aside from the collapse of the Soviet Union in the early 1990s, but not in the 1980s, what changes would justify such a dramatic shift? In our specific concern for agriculture and the conditions of the rural poor, we see that, far from fundamental changes, there has been little basic change, and that the institutional obstacles to agricultural/rural development have even worsened. We have already demonstrated the persistent and increasingly rural poverty and malnutrition, and we shall see in Chapter 3 that land concentration actually *worsened* in many LDCs. Institutional obstacles to growth and technical change in the declining food-producing sector also continue. The number of the poor and malnourished has increased considerably despite the World Bank and International Monetary Fund (IMF)-induced economic reforms for enhancing economic growth, exports and trade globalization.

In this deteriorating tendency in poverty reduction, it is also disappointing to see a downward trend in the volume of international aid to LDCs in favour of the former communist countries of Eastern Europe and those that were members of the former Yugoslavia and Soviet Union (for example Hungary, Romania, Georgia, Bosnia, Macedonia, Slovenia, Estonia and Poland). In 2001, these countries received double or three times the amount of international assistance (Overseas Development Agency, London) per capita compared with such needy countries with persistently high rural poverty as Sudan, Congo, Ghana, Guatemala, Colombia and Yemen, just to mention a few (UNDP 2003: Table 16).

The likely effect of this international regressive response to persisting poverty and hunger and, in turn, on agricultural output growth, workforce productivity and, in particular, food production will be examined briefly in this section but fully in Chapter 3.

Ambiguities in definition and measurement

Let us start with the two internationally committed goals of eliminating hunger by 2000 and halving poverty and undernourishment by 2015. Monitoring progress in both suffers from ambiguities: what is to reduce or to eliminate? How to measure change? And what is the reference year for comparison?

The 145 governments attending the 1979 WCARRD Conference in Rome agreed 'to eliminate' undernutrition by the year 2000, meaning the removal of every single undernutritional case because of the world leaders' 'belief that poverty, hunger and malnutrition retard national development efforts and negatively affect world social and pro economic stability' (WCARRD's Declaration of Principles No. 7). In addition, they agreed that governments should: 'impose ceilings on the size of private holdings', 'giving precedence to the distribution of acquired land to established tenants, small holders and landless agricultural workers' and 'fix specific targets for the reduction of rural poverty' [WCARRD Program of Action 1. A (I, ii, iii and iv) and II, A (I, ii, iii)].

Because of these generalities and ambiguities, I was given the responsibility within FAO, supported by a very able team of statisticians, nutritionists and economists (for a period of 6 years, 1980–85), to assist governments in the establishment of quantitative indicators for monitoring progress made in the reduction of poverty and undernutrition in rural areas and inequality of agricultural land distribution, against the situation in the base year 1980 (see El-Ghonemy 1984a; FAO 1988). This systematic work has progressed until the early 1990s when the governing body of FAO – under the influence of the World Bank and the USA – decided to terminate it, saying 'economic liberalization has the potential to act as a vehicle for rural poverty alleviation and its eventual eradication' (Conference C95/INF/22 1995: 7). Consequently, the phrases 'land reform', 'agrarian reform' and 'monitoring rural poverty' were abolished in the FAO administrative structure, and the team on monitoring progress in the alleviation of rural poverty disbanded. Sadly, we should recall that, in 1979–80, FAO was mandated by the international community to the lead role in these subjects within the entire United Nations system. In Chapter 2, I shall return to examine this shift, and to understand its foundation and explore its policy implications in Chapters 7 and 8.

In monitoring undernourishment reduction, we have seen the sweeping WCARRD (1979) goal of its 'eradication' by 2000 but, in other internationally agreed goals, there are other deficiencies. Although the World Food Summit of 1996 has set the goal of halving the *number* of the undernourished by 2015, the World Summit on Social Development (Copenhagen, 1995) established another sweeping goal of eradicating poverty and eliminating undernutrition, without setting a time frame in its Plan of Action. Also, the Development Assistance Committee (DAC) of the OECD (Paris, 1996) set the goal of halving the proportion, not the numbers, by 2015. Even the 2001 UN General Assembly millennium development goals define the first and second goals of halving the proportion of poverty and the people suffering from hunger between 1990 and 2015 (see UNDP Human Development Report 2003: Ch. 1). There is also no uniformity of poverty measurement; some use the World Bank's 1 or 2 dollar a day poverty line, whereas others use a country-specific household income/expenditure poverty line calculated for commodities consumed and prices in rural areas.

Can poverty and undernutrition be halved by 2015?

The available evidence suggests that the 50 per cent reduction depends on the poverty measurement and base year used and the accuracy of calculated national averages of daily calorie supply per person. It also depends on the assumptions made about rural population growth/life expectancy, fertility change and rural–urban migration. Besides, the projection depends upon what would happen to total economic and agricultural growth as well as income distribution in the heavily populated countries (China, India, Indonesia, Brazil, Nigeria, Bangladesh and Vietnam). Furthermore, available information indicates the importance of LDC governments' improvement of their national statistical capacity to build data required for measuring progress and their choice of pro-poverty-reducing policies combined with the volume and effectiveness of both domestic resources allocation and foreign aid to agriculture, particularly to the food subsector. This agricultural food focus is crucial for the targeted reduction, especially in the South Asian and sub-Saharan regions where rural poverty and undernourishment are concentrated.

The United Nations Task Force on Hunger, an international group of experts on these subjects monitoring progress towards meeting the millennium goal, has found that, of the 815 million people suffering chronic undernutrition in 2000–02, the figure was only 9 million fewer in 2004 (i.e. an average annual reduction of 0.7 million) and that 80 per cent of the total are rural and depend on agriculture for a living. They consist of 50 per cent small farmers, 20 per cent agricultural landless wage workers and 10 per cent pastoralists (FAO 2004a: 25, Table 1). Given these several estimates, one can say that halving the undernourished to 407 million in 2015 is out of reach of most developing countries. For the benefit of the reader, Appendix B summarizes the millennium development goal on cutting poverty and hunger by half in 2015 and on related arguments.

With regard to halving the total number of poor people in developing countries from 1.27 billion in 1990 to 0.75 billion in 2015, using the international simplified standard measurement of the poverty line (1 dollar a day per person), the World Bank projects that the 2015 proportionate target is likely to be achieved (World Bank 2001: Table 1.8). This is because great progress has already been made in China, India, Latin America, North Africa and the Middle East, whereas in sub-Saharan Africa, the *proportion* will remain high, declining from 47.7 per cent in 1990 to 39.3 per cent in 2015, and the *number* of the poor will increase from 242 million persons in 1990 to 345 million in 2015. The number will remain very high for several decades to come depending on the frequency of natural disasters, armed conflicts and civil unrest. In all developing countries, excluding East European and other transition countries, it is projected that the number will fall from 1,269 million in 1990 to 749 million in 2015 (i.e. 60 per cent of the 1990 level and not half as internationally targeted). See similar bleak prospects in Naschold (2004: 116–18).

Before we end this section, two remarks need to be made. First, according to projected population data for 2015, rural population continues to be high at nearly 37–40 per cent or an approximate order of magnitude of 280 million persons in LDCs' rural areas. Second, although the World Bank's international poverty line of $1 a day per person is useful and simple in drawing public attention to a complex development problem, it lacks country-specific variations in: food consumption by cultural norms; rural–urban food prices that differ from both national averages and the national cost of living index. Consequently, variations exist in calculating the purchasing power parity with the US dollar against 1993 prices. Furthermore, the World Bank's poverty reduction projection (1990–2015) assumes high average annual per person GNP growth of between 2.5 and 5.5 per cent (see Srinivasan 2000; Deaton 2001; FAO 2003a: 215–16).

Does international aid reduce rural poverty?

I have examined elsewhere (El-Ghonemy 1984b: 45–6) whether international aid has effectively supplemented domestic resources for financing poverty-reducing agriculture and rural development, especially in low-income, food-deficit countries. Total aid in 1981–82 was far below the target of 0.7 per cent of donors' GNP, falling to 0.37 per cent, and food aid to these countries fell in physical terms to 79 per cent from 81 per cent of total shipment in 1981–82 average terms, contrary to what was committed at the 1976 World Food Conference. Besides, there was resentment among some recipient countries about the political intervention of donors, using aid as a political instrument to serve their own strategic interests. Targeting aid to poor countries and targeting the poor within each recipient country was, and still is, the primary concern over the effectiveness of aid in reducing poverty.

It is indeed deplorable if total aid, particularly food aid, to developing countries does not reach the undernourished poor in rural areas. In their careful assessment of aid, Robert Cassen and his associates remarked in 1986 that aid has been ill-directed to the poor: 'very little of it, has been directed at, or has had any impact, positive or negative, on the poorest (the 10 per cent at the bottom of the income distribution' (Cassen *et al.* 1986: 110). They report a number of conflicting views revealed by an evaluation of food aid (e.g. food supplied is allegedly sold off or taken by the army and civil servants; Cassen *et al.* 1986: 161).

The subject of poverty-reducing aid was recently addressed by IFAD in its *Rural Poverty Report*, 2001: 40–2). In its survey of 47 developing countries, IFAD revealed that 'aid is very badly targeted on countries with high extreme poverty'. The criteria were the higher proportion of rural population and of the rural poor, where the proportion of persons engaged mainly in agriculture was 64 per cent in sub-Saharan Africa and 55 per cent in South Asia but only 21 per cent in Latin America (p. 41). 'If properly targeted total aid would have covered 37 per cent of the total one-dollar-a-day poor in the sample survey year, 1995' (p. 40).

A new approach to the study of the effect of aid on poverty reduction is by way of relating the percentage of total aid received to the GDP of each country and to its economic growth (GDP per capita) as well as to the poverty level based on the US$2 a day per person poverty line. Employing data from 62 developing countries, from 1974–77 to 1994–97, Collier and Dollar (2001: 1787–1808) found that aid has a positive effect on economic growth and poverty reduction and that this effect is *conditional* on the policy quality (appropriate policy and economic institutions). They consider that the policies of South Asian countries (India, Bangladesh, Pakistan and Sri Lanka) are the nearest to their hypothetical model. As we shall see in Chapters 5 and 6, these countries have followed piecemeal, partial, redistributive land reforms. We should note the ambiguity of the term 'policy quality' and also recall the influence of donor countries' own political preference and commercial interests on recipient countries. I shall examine this point in Chapter 7 with regard to market liberalization and government intervention for land redistribution, and in Chapter 8 about the consequential variation in the pace of rural poverty reduction.

2 Reform of land property rights between the state and the market

Principles and concepts

In the introductory chapter, I have repeatedly mentioned several terms and phrases having a wide range of meanings in their usage by different people that tend to make their interpretation controversial, *without* investigating their relevant specification to poverty-reducing mechanisms of secure land access. Examples of these expressions requiring specification are: state interventionist authority for a speedy poverty reduction by redistributive land/agrarian reform; morality of unjust land acquisition combined with exploitation and institutional monopoly in land and labour markets; re-emergence of neoliberalism for enhancing equity and efficiency that can only be satisfied by a private market economy free of state control; problems of comparability with regard to measurement of poverty, inequality, malnourishment and land concentration, and so on. To clarify the meaning of these fundamental terms and phrases before we move to empiricism in the following chapters is the task of this chapter. It also attempts to set the frame of reference for the analysis of countries' experience in the rest of the book. Otherwise, the discussion enters a vicious circle of philosophical puzzles about the different meanings of one term and seeking answers to what is meant?

The discussion consists of five major sections. It begins with the nature of property rights and duties in land and their several land tenure arrangements. The second section defines institution and discusses a wide range of institutional forms of property rights, with emphasis on institutional monopoly and its serious effects on income distribution and poverty in rural areas. The section also discusses non-market arrangements: inheritance, bequest; intermarriage land property transfer; and voluntary non-government distributional arrangements suggested by Rawls (1972) in his *Theory of Justice*. The third section discusses the foundation of the post-1980 ideological shift towards the restated neoclassical principles behind neoliberalism in the context of the distribution/transfer of property rights. These principles have a cluster of ideas with regard to limiting state authority in a market economy. The rest of the chapter deals with the key concepts with different meanings, different usage and measurement problems. They are poverty, malnutrition, food insecurity, land and income distribution, social justice and the complementarity between secure access to land and rural development.

The nature of land tenure and property rights

The institution of property rights in land is the heart of land tenure arrangements and their alternative forms of redistribution. By property rights is meant private or public property institution that is legally protected by excluding others or outsiders from its sale, lease and use without consent. In the western system of thought and historical tradition, the institution of *private* property rights is a foundation of capitalism, and is considered to be essential to democracy. In pure socialism, the abolition of private property is an integral part of the entire economic system for attaining rural development via central planning and collective management/ownership with allowance for individual cultivation of food crops required for household consumption. But, we suppose that, whatever the political philosophy underlying the institution of property, there is no reason why an incentive-driven privately owned family farm or *communally* owned land cannot enhance rural development. Under both arrangements, the state adjusts, in varying degrees, the institution of property rights in land according to public interest.

Adjustment of exchangeable rights in land property

In 1890, Alfred Marshall, the founder of modern economics, wrote: 'Taking it for granted that a more equal distribution of wealth is to be desired, how far would this justify changes in the institution of property, or limitation of free enterprise?' The institution of property in the sense used by Marshall is taken to mean the intangible or the exchangeable rights in ownership and use of property as determined by law or custom. This content of property is distinct from physical or corporeal property [to use the terminology of MacLeod (1867) and Commons (1923, 1934)]. It denotes leasing, sharing arrangements, indebtedness and mortgage, inheritance and mortgage regulations for the transfer of property rights from one generation to the next. This exchangeable content of property rights also denotes security of tenure and property-based power or political advantages, i.e. a landless poor is politically dominated by landlords and have no power in community decision-making. In addition, this content of property rights determines both the flow of accrued income and its distribution among the participants through their interaction with market transactions (e.g. Adam Smith's famous phrase 'the invisible hand' that stimulates production and distributes gains between participants in the exchange process). Also, the exchangeable rights induce or inhibit investment in improving the productivity of the physical content of property by way of technical change (e.g. irrigation and soil improvement by applying fertilizer) and, therefore, affect the intensity of land and labour use.

If our interpretation of Marshall's notion is correct, the question is: should these rights in private property be preserved on the grounds that property and economic freedom are sacrosanct irrespective of their distributional effect, or should they be conditioned by the state power with a view to maximize social welfare? As Galbraith remarks: 'As long as it remains in private hands no others can possess power'. In non-socialist doctrine, in contrast, private property is so important as a source of private power that it cannot be concentrated in the hands of the government, yet it should enjoy the general protection of the state. There remains the question of 'how extensively the State should intervene to get a wider distribution of property (and associated income) and thus of the power emanating therefrom' (1984: 47, 87).[1]

This question has engaged the interest of many philosophers and analysts from different strands of economics. In his *Wealth of Nations* (1776), the prudent Adam Smith conceived a principle governing the role of the state 'Protecting, as far as possible, every member of the society from the injustice or oppression of every other member of it' (Book IV, Ch. IX: 651). This principle implies that the state exercises its political power to restrain the economic freedom of individuals or corporations who abuse such freedom for attaining private gains at the expense of others and social gains, as in the case of violating property rights by monopolists and leaving a part of the arable land area unused. In his scholarly conducted argument, Baumol says: 'I believe that the politician is, in many cases, justified in taking, and indeed forced to take, action on many (practical) problems: perfect analysis or no (1965, Part II: 204–7). In fact, this is what governments are doing.

Private property, despite being the central bond of capitalism, has aroused public concern over the consequences of institutional monopoly manifested in the concentration of landownership in a few hands combined with growing landlessness, chronic indebtedness of the peasants and eviction of tenants. These manifestations are among the proximate causes of persistent rural poverty, and they can threaten political stability. When the balance of power swings towards the interests of the poor peasants and the landless workers, the state intervenes to condition the institution of property rights and, in varying degrees, to limit the economic freedom of entrepreneurs in agriculture. In this quest for justice, the state in a capitalist system does *not* abolish private property in land but, instead, it regulates ownership rights and rectifies factor market defects in the rural economy. The state, or its executive branch, the government, supports the operation of the market mechanism (e.g. land and credit markets) in many ways, including the issue of laws to protect property rights, and fulfils business ethics in land mortgage as well as banking and trade operations.

The extent of adjusting the role of the market

The extent of state intervention in private property market economies is unlimited, expressed by the 1974 Nobel Laureate in Economics, Friedrich von Hayek, as 'all governments affect the relative position of different people and, that there is under any system scarcely an aspect of our lives which might not be affected by government action' (Hayek 1978: 81). This is certainly true. In so far as government does anything at all, 'its action will always have some effect on who gets what; when and how'.[2] In socialist economies, on the other hand, the government's nationalization of landed property and non-labour means of production in agriculture is combined with central planning. This combination severely diminishes the role of the market in determining the distribution of income in the entire economy. The alternative to the market is centrally designed distribution of wealth and income by a highly statist control of production, exchange and distribution. This coherent system does not provide absolute equality, but it does attempt to ensure minimum inequality. As an ideological preference, the historical experience of China, Cuba, Hungary, Mexico and Russia shows that this approach was a radical response to long-established feudal systems in agriculture. In both economic systems, well-defined property rights, law enforcement and regulatory apparatus are required by state intervention.

These regulatory arrangements are necessary for the national development aims of welfare, social justice and sustainable economic efficiency. In Chapter 8, I shall return to this central question of interdependence or complementarity between government intervention in land and credit markets and their workable institutional arrangements for reducing poverty and increasing productivity.

The meaning of institutional monopoly and exploitation in agrarian economies

At this point in the discussion on landed property rights, the concepts of institutional monopoly, exploitation, feudal relations and land concentration need to be explained within agrarian economies. This refers to economies in which the influence of the land tenure system's linkage with agricultural credit supply, production, employment and income distribution is dominant in the national economy.

What is meant by institutions?

The combination of institutions and relevant principles of economics enables us to study the elements of monopoly/monopsony powers in agricultural factor markets (land, labour and capital) and their poverty and distributional consequences in rural areas. Institutional determinants of these socio-economic welfare components have received less attention than they deserve and, despite Alfred Marshall's realization in his *Principles of Economics* more than a century ago that institutions are important and his proposal that the study of land tenure and related institutions of property rights are questions to be investigated by economists (1890: Book 1, Ch. IV), it is only recently and since the work of John Commons (1934) and Roland Coase (1960) on transaction costs that the subject of institutions and institutional change in the agrarian economy has received the economists' deserved attention and recognition, by the awarding of the Nobel Prize in Economics between 1973 and 2003 to ten eminent scholars for their contribution to the importance of institutions in understanding economic performance.[3]

In simple terms, institutions mean the set or bundle of rules and regulations governing property rights, factor market transactions and price information systems, including asymmetric information (i.e. information possessed by only one side in a market transaction, buyer or seller, as for instance that between peasants and landowners and moneylenders). They also mean law enforcement and bureaucracy and fraud in trade, banking and investment activities as well as regulatory apparatus in taxation and judicial systems. They compose a chain that is interlinked. For the purposes of our discussion, four elements of institutional monopoly that are termed 'institutional constraints or obstacles' to agricultural/rural development in some literature are examined separately as follows.

Land concentration (LC)

As productive land is the crucial income-yielding, food-producing and labour-using asset in agriculture, the concentration of its ownership is the principal determinant of social power, political advantages and other monopoly elements. For example, it leads to the control of the terms set for renting land out to peasants, determination of wage levels, the number of workers hired, the share in purchased inputs and in the sale of farm output. The extent of such monopoly power depends on the size and locality of the large farm and commercial plantation. It is also determined by the demand for renting the land and the supply elasticity of labour. If, for example, the landlord or plantation manager is the sole rentier of land and buyer of hired landless workers in his locality (due to the substantial size of his farm), he or she can gain supernormal profit from charging high rents and simultaneously paying wage rates lower than those that would apply under a competitive labour market. Thus, the combination of the farm size and its share in the total landholdings in a specific location is a significant characteristic of market power in agrarian economies.

To understand the relation between concentration of landholding and poverty, we need to analyse how LC limits the options of the poor to raise their earnings and nutritional standards above the poverty line. These option limitations are manifested in seriously restricted access to the rent or purchase of a piece of productive land, in chronic indebtedness to landowners and in the very low return on labour. This low level of earning, and its corresponding low level of food consumption combined, form a major determinant of poverty.

Transaction costs and barriers to entry into land, labour and credit markets: localized poverty trap

Transaction costs are those incurred in surmounting barriers in trading goods, registering land, securing a written lease contract, enforcement of laws and the four dominant barriers to entry into factor markets explained below. Habitually, these costs are neglected by many economists and accountants despite their considerable amounts. One would expect that these transaction costs are disproportionately higher for the mass of illiterate small farmers than for large and rich landowners who have the knowledge, contacts and influence on bureaucracy and credit bank staff to overcome these costs and to obtain production inputs and the amount of credit needed faster and more cheaply. But, before outlining these four barriers to entry in agrarian systems, we briefly distinguish between them and those in industry.

In defective land tenure systems characterized by land concentration, the insidious conduct of agrarian entrepreneurs differs from that practised in manufacturing. Practices such as commercial advertising, licensing, trademarks and patents as a means for property rights' differentiation of products, price slashing by new entrants or mergers designed to expand output and reduce cost are not applicable to agrarian entrepreneurs (see Demsetz 1982), nor is anti-trust or monopolies and restrictive trade legislation applied to land tenure and the related market power in agriculture. In the absence of enforced regulatory legislation on tenancy arrangements or restrictions of property rights in land, landlords, multinational plantation managers, sellers of pump irrigation water (waterlords), farm equipment dealers and moneylenders practise malicious monopoly powers without fear of legal penalization.

The first dominant practice barring entry into the land market is the auctioneering of land, with tenancy middlemen who can meet the financial requirements that serve as an insurance against risk. These middlemen subdivide the land into smaller units, subleasing them to small tenants at rates higher than those paid to the absentee landlord or his agent. This practice provides a high profit margin. The second practice is the hierarchical system of contractual and subcontractual hiring of large numbers of landless workers. These workers are paid wages far below their average and marginal productivity in order to provide a profit for each level of their hierarchy. The third dominant practice is rationing agricultural credit by barring tenants, small owners and landless workers from entering the capital market. By restricting access to credit, a large group of peasants is denied investment for higher productivity, and pays the opportunity cost of possible increases in their earnings. The results are a higher cost borne by peasants than that of medium and large farmers having easy access to credit. The poor peasants' added costs take the form of higher interest rates paid to moneylenders, usually linked with the commitment of the borrowers to sell their crops to them at reduced prices. The fourth barrier to entry is in denying agricultural workers their right to organize trade unions and in restricting activities such as the right to strike.

The multiple functions of large farmowners give them exclusive and special monopoly powers. Imagine a rural locality, in which a big landowner is the mayor of the village, a trader, a moneylender, an owner of water pumps for irrigation where water is scarce and, in addition, he may be an influential politician. If he or she is the sole rentier of land, as well as the sole employer of landless workers, he or she is a capitalist with absolute monopolistic–monopsonistic power in his or her locality forming a localized poverty trap.

Special monopoly advantages of multinationals operating in
agriculture

As powerful institutions in the production and trade of high-value crops in less developed countries (LDCs) and at international level, multinational corporations (MNCs) demand special emphasis. They have an increasingly important influence in shaping the agricultural economy of many developing countries via private foreign capital investment, transfer of technology and management skills, net capital outflow and control of some economically strategic export crops. Our concern is limited to the influence of MNCs' operations in agriculture on the pattern of distribution of land, income/consumption and power, labour utilization and the reallocation of scarce resources between food and non-food crops. Yet another related area of concern is identifying the MNCs' elements of oligopolist market power, including the collusive alliance with national agents in developing countries. All are questions that suffer from a scarcity of hard evidence.

By virtue of their very nature of seeking high profit and their substantial size and mode of negotiating contracts with developing countries, multinationals are quite able to surmount barriers to entry into holding land for production, processing and export of economically strategic crops. They also appear to be quite capable of securing high rates of monopoly profit and generous concessions and preferential treatment in taxation, in pricing and in repatriating their high monopoly profit (usually tax free) to parent firms in their home countries. The special advantages of MNCs lie in their monopolistic package, which includes the integrated system of research capabilities, crops and technology choice, timely input supply, processing, sophisticated marketing techniques, transport, export and high level of skills in supervision, management and organization within a gigantic operation.

To maximize the economic reward of high profits from these monopolistic/ oligopolistic advantages, while minimizing the risk, MNCs use their *own* government's political facilities, obtaining sufficient support for being part of their government's foreign investment policy. To obtain the necessary political lobby and social cover as an insurance policy against potential risk, MNCs seek alliance with influential landlords, businessmen, senior government officials and politicians in host countries. To further reduce risk, MNCs develop safeguards against possible future nationalization or procure guarantees for full compensation payment. Towards this end, they gradually switch from direct investment in plantations to joint ventures, contract farming, processing agricultural raw materials, marketing and management of public corporations producing high-value crops. MNCs also use their influence on governments intending to issue land reforms. In this case, MNCs strive to exempt their plantations from application of the size ceiling on private land property or to receive a guarantee for absolute exemption (as occurred in the Philippines' Land Reform Decree No. 27 of 1972).

Governments of developing countries, on the other hand, are usually starved of capital needs. They cannot resist the prospects of advances in technology and management of the modernization of the export crop sector to earn foreign exchange. With this goal in mind, it is not surprising that developing countries provide MNCs with generous incentives and preferential treatment. In many cases, LDCs' governments are unfamiliar with the internal structures of MNCs, the methods employed to calculate profits and manipulate prices (transfer pricing) of imported capital goods and exported products between the MNCs' own subsidiaries and firms located in other countries. Nor can they interpret the twisted accounting systems of MNCs sufficiently to estimate costs and the exceedingly difficult but highly important distribution of gains in terms of net flow of capital, net contribution to export earnings, public revenue, and the volume and share of labour benefits.

Exploitation as an element of institutional monopoly

In my country-specific field studies and discussions with several senior officials, I have found that governments assign high priority in redistributive land reform aims to abolishing social injustice manifested in the exploitative elements of institutional monopoly in their agrarian economies, although they seldom use this terminology. Instead, they view institutional monopoly as 'feudalism', the term borrowed from pre-seventeenth-century Europe. Feudalism in its *real* sense (with coercion and military service to landlords), however, did not exist in the pre-reform situations of contemporary developing countries. Country experiences suggest that governments carrying out land reform express the presence of exploitation in terms of a combination of the following institutional monopoly forms:

1   landownership concentration;
2   serfdom in terms of bonded labour and landlords exacting illegal levies and services from their peasants;
3   the dominance of foreign-owned land in agriculture and the owners' collusion with government administration to depress the earnings of tenants and landless workers;
4   widespread absenteeism among private owners of large farms associated with renting out land under insecure tenancy, at high rents and without written contracts and compensation payment for eviction or land improvements;
5   the heavy burden of the peasants' indebtedness;
6   illegal land grabbing.

Exploitation, although a familiar expression widely used by countries justifying land reforms, is nevertheless ambiguous, and so more questions must be raised. For example, what is the mode of market power relations in the use and exchange of resources which could be legitimately viewed as exploitation? What criteria can we use for the identification of the exploiters and the exploited? Based on his compiled data from Irish and English agriculture during the period 1851–71, Karl Marx conceived exploitation in terms of the capitalist extraction and accumulation of 'surplus value', particularly from displacing labour by the newly invented machinery of his time, and from setting wages at socially determined subsistence. He considered the rate of surplus value as a measure of exploitation. From the class conflicts within capitalist agriculture, he identified the capital owners as the exploiters (e.g. landlords and the spinners of cotton and wool) and the labourers as the exploited living in 'increased misery'. The prevailing exploitative relations in production, in his words, 'have sprung up historically and stamp the labourer as the direct means of creating surplus value' (1906: Part V, p. 558).

Another group of exploiters was added by Lenin in his study of late nineteenth-century Russian rural economy. In that work, *The Development of Capitalism in*

Russia (1899), Lenin considered the middle farmers with a commercial orientation of their means of production as exploiters of the poor peasants who own *only* had their labour power and working animals. The identification of the exploitation criteria in production and exchange was first attempted during the 1920s by two groups of Russian scholars led by Chayanov and Kritsman. In his comprehensive review of their work, Terry Cox (1986) reported a number of indices that were used as criteria in the identification of exploitation relations. Another explanation of exploitation comes from two American scholars, Nozick (1974) and Roemer (1982). They go beyond the Marxian hypothesis, both attacking his explanation. From their point of view, exploitation is seen as a violation of private property rights and entitlements. This view is, in a sense, an expansion of the ideas founded by John Locke in the seventeenth century with regard to private property rights, including free labour and its products. To Nozick, exploitation is not the rate of appropriation of surplus value, which he and Roemer consider applicable to feudal conditions. Each places different emphasis. Nozick argues that the exploited do not possess the scarce entrepreneurial abilities and marketing skills to innovate and to bear market risks and uncertainties in free market transactions based on voluntary exchange. Thus, his argument is extracted from a competitive market mechanism that does not characterize the rural economy in many developing countries that have, since the 1950s, instituted redistributive land reform.

All these contributions to the meaning of exploitation are formed by individual scholars with varying backgrounds and within each author's unique system of analysis. Abstracted from different historical–institutional conditions, the analytical reasoning behind each interpretation seems to be consistent with each conclusion reached. For example, John Locke, the founder of the concept of property relations, was arguing against the absolute power of arbitrary 'divine rights' of the King of Britain in the seventeenth century. Under his rule, the King granted absolute monopoly rights in land to a few noblemen who were entitled to subdue their landless workers and peasants. Despite the unique analytical reasoning behind each interpretation of exploitation, I find a number of common elements. Locke's notion of exploited property rights in granting and inheriting land are narrowly conceived by Marx, corresponding only to labour and what it produces as the substance of its value. Marx's appropriated surplus value corresponds, in part, to the classical economic rent or monopoly profit resulting from the monopolist's deliberate wage-setting below the value of the worker's marginal product in an imperfect market.

Non-market land property transfer: inheritance rules

Inheritance, bequest and interfamily marriage are the most common channels for landownership transfer in developing countries, except for the poor landless agricultural workers whose families are too poor to own land or to marry into the landed class. These key non-market instruments for intergenerational property transfer have been verified.[4] Because of their widespread practice in land transfer and owing to existing variations in God-given inheritance rules, a summary of these moral principles by the major world religions is presented for the benefit of the reader.

Land inheritance in Christian and Jewish rules

In Christianity, inheritance rules differ widely from those of Judaism and Islam. The Christian tradition reveals a great deal of diversity. It ranges from passing the property undivided to a single heir (usually the first-born child or the surviving spouse, an arrangement known as the right of primogeniture) or allowing at least one-third of the property for the surviving spouse or for charitable purposes. In the Protestant tradition, inheritance expressed in a written will permits placing the property after the owner's death at the disposal of whomsoever the owner pleases and on whatever terms, thus perpetuating inequality in wealth ownership. In a sense, this practice violates the western capitalist ideal of 'to each according to his or her productivity': this has led an American philosopher to call for the abolition of inheritance.[5]

Judaism, on the other hand, generally defines the distribution of property among the legitimate heirs according to specified shares. I say 'generally' because there is flexibility in Judaism between writing or not writing a will (*katub*); passing the entire property to a single heir; deciding on the distribution of property of an intestate person by means of the rabbinical courts or giving freedom of testament that can disinherit one or more eligible heirs. In setting out the order of precedence of the heirs, the Torah rules that the eldest son inherits double the share of the others. However, a father can make a will whereby he empowers one or more of his inheritors to acquire his whole estate, thereby denying the others their share. Justice in inheritance is provided by the early rabbis' rulings that brothers were obliged to provide fully for the needs of their sisters, even if they were thereby reduced to poverty themselves.[6] Should a person wish not to follow this divine mandate, he or she can provide the means of support as a gift. In Israel, however, Jewish people maintain a high level of observance, especially within the Orthodox and conservative section of the population.

Islamic inheritance rules

Rules of private property inheritance in Islam are mandatory, laid down in precise terms in *Surat al-Nissa* (11–13) and *al-Baqara* (241) of the Qur'an. Subsequently, the rules were explained for practical application by the Prophet in Hadith. Only the subdivision of indebted property and the shares of non-primary inheritors were differently interpreted by jurists (Abu-Zahra 1963: 25). Known in Arabic as *mirath*, the principles of inheritance guarantee the right of legitimate inheritors, particularly minors, and also of creditors. The principles also give the estate owner the option to allocate, bequest a *heba*, after death up to a maximum of one-third of his or her private property for charitable purposes (*wassiya*), but not to legitimate inheritors (*la wassiya li wareeth*).[7] The balance constitutes the compulsory inheritance (*forud al-tarikah*) to be subdivided in clearly stipulated shares among entitled heirs, after funeral expenses and any debts owed to creditors have been met.

The first and fundamental principle is the mandatory subdivision of the inherited estate into shares, *nasseeb*, i.e. one-half, one-quarter, etc., whereby the male's share is double that of a female heir. The second is *takharoj*, in which the physical splitting of inherited land is not mandatory, whereby it can be retained and managed – in a single production unit – by one of the heirs who is trusted. Thus, the size distribution of property rights from one generation to the next and the resulting income distribution are influenced by population growth and family size. Importantly, the property owner's longevity (length of life) is also crucial: the longer he or she lives, the slower the property transfer among heirs and the longer the cycle in the transfer of property rights and the income distribution effects. In fact, the role of inheritance and bequest in the process of land accumulation and intergenerational transfers related to long-term distribution of income (and savings) has recently received increasing attention from economic analysts, particularly the 1985 Nobel Laureate for Economics, Franco Modigliani (1975, 1988). Phelps Brown (1988) also estimated the share of inheritance and inter-family bequests at one-half of USA and UK national wealth.

On land reform and agrarian reform

It may be odd, if not absurd, to defer the inquiry into the meaning of this key concept to this point in our inquiry. Simply, its deferred discussion is for sequential reasoning, that is after we understand the meaning of property rights and duties in land and related institutional arrangements. Let me introduce first the state redistributive land reform followed by the voluntary market-based land reform.

In general terms, land reform is a public action assigning a specific role to land tenure to amend what are considered by the state to be iniquitous practices against the public interest – practices that create conditions inhibiting poverty-reducing rural development. In concrete terms, land reform is conceived in this study as a redistribution of private land property and use rights under different institutional arrangements enforceable by law. The aim is to remove the already identified barriers to entry into the factor markets (land, labour and capital or credit) and to provide peasants with command over their food needs, thereby rapidly reducing undernourishment, absolute poverty and inequalities. However, this does not mean that *any* land reform can rapidly realize these changes. Nor does it mean that land redistribution *alone* can achieve these results and sustain them over time. Although land reform is a strong demonstration of political commitment directed to abolishing exploitation, empowering the poor household beneficiaries and directly attacking rural poverty, various redistributive land reforms differ in their aims, pace and scale of implementation, as to results. The principal determinants of these differences are as follows:

1 political commitment and unbiased bureaucracy;
2 scope of change – the scale and terms of redistribution, particularly the country-specific levels of size ceiling on the ownership of land and the minimum size of distributed units in relation to the extent of landless peasants;
3 implementation capability of state institutions, including the administrative capacity for enforcement of legal provisions and whether these provisions contain loopholes and ambiguous rules;
4 complementarity of other public actions and institutional arrangements, particularly the registration of landownership titles and lease contracts and the timely supply of production inputs;
5 pace of implementation without uncertainty.

Irrespective of ideological differences, the performance of land reform programmes is differentiated by a chain of these five elements. Without exception, the effects are functionally dependent upon each of them, that is they form links in the composite function of land reform. In some literature, particularly that concerning the Latin American region, the term 'agrarian reform' is used instead of 'land reform'. The term in Spanish embraces a wide range of public institutional changes in agriculture which may – but do not necessarily – include the redistribution of private landed property. Agrarian reform usually refers to land settlement schemes in publicly owned land, land registration, lending institutional credit to tenants to purchase land in the open market, consolidation of fragmented holdings, regulation of tenancy arrangements, and so on. If land policy leaves the existing skewed distribution of privately owned land and rural socio-political power unchanged, these institutional measures cannot be considered to be land reform under our definition. This is because:

• they represent an evasion of the central issue of inequality in landownership and institutional monopoly;
• in the face of existing political power of landlords over government bureaucracy, such changes, particularly the tenancy regulations, cannot be effectively enforced; and
• with tenancy regulation, as with scattered land settlement schemes, social injustice, agrarian exploitative relationships and the prevalence of rural poverty are not likely to be significantly reduced where the concentration of landownership and related political power is high and where the pattern of agricultural GDP growth perpetuates inequality.

According to our definition, genuine land reform could be termed agrarian reform, but the converse need not be true. The term 'agrarian reform' has been expanded to an extent rightly noted by three seasoned scholars. Kenneth Parsons (in El-Ghonemy 1984b) describes it as 'reformation of the structure of the agricultural economy', whereas, according to Doreen Warriner (1969), 'its use in this wide sense blurs the real issue. The net is spread wide, the catch is miscellaneous.' The third eminent scholar is Michael Lipton, who asserts that:

It is absurd to exclude major methods of land redistribution, whether dis-
tributivist or collectivist, from the definition of genuine land reform, for
both involve radical equalizing changes in the land-based structure of rural
power. Not only is such redefinition poor logic; the attitude behind it plays
into the hands of landlord-politicians. Perhaps the most common of all these
over-rigorous evasions is, 'don't do anything till you can do everything, so
do nothing . . .' This broad definition lets 'the rich farmers keep their land,
fertilized with the crocodile tears of frustrated reformers'.

Lipton (1974: 274–5)

The importance of post-land reform sustaining gains

With land reform so defined, as with agrarian reform, the reallocation of property
rights is not sufficient to attain potential gains. Nor does it guarantee a continu-
ation of the beneficiaries' incremental income in real terms, or their command
over their household food needs. Other complementary inputs and measures are
a prerequisite for sustaining initial gains over a long period and reinforcing their
linkage with the process of rural development.

The speed in replacing the production and marketing functions of former
landlords is crucial. These arrangements are instrumental in ensuring the supply
of production inputs such as water, seeds, fertilizers and the sale of marketed
surplus. Furthermore, investment in improving irrigation, particularly in arid and
semi-arid areas, is important. The shorter the lapse of time between abolishing
the old institutions and their replacement by a relevant new order, the less the
risk of destabilizing the flow of income to the beneficiaries and marketed surplus
to urban centres. Any weakness or delay in these logistical operations is bound
to lead to a short-term output and income fluctuation, which is usually used by
opponents of land reform to strengthen their argument against it.

Gains in the economy as a whole are dependent upon the scope of land reform (i.e. the extent of land property redistribution) and its impact on agricultural
growth. As development records show a positive relation between agricultural and
national GDP rates of growth, sustained agricultural output growth after large-
scale land reform is essential. It follows that raising output in the reformed sector
requires more than the celebrated remark by Arthur Young, 'the magic of property
turns sand into gold'. This 'magic' works only with the motivation and energy
invested by the beneficiaries combined with an adequate public investment and
supply of complementary inputs.

As taxpayers and food consumers, the rest of the nation cannot help but be interested in what happens in the countryside. The urban sectors and agro-industry
desire a stable flow of agricultural products, particularly food crops and animal
products. As land reform is not without cost, there are financial considerations as
well, in addition to the transaction costs discussed earlier. Speedy implementation and complementary inputs to sustain its benefits incur expenditures from the
national budget (and in turn the taxpayers). The scale of the burden on the national
budget depends upon the pre-reform conditions, the extent of land redistribution,

the infrastructural investment required, the terms of compensation to be paid for expropriated property (particularly if it was owned by foreigners) and whether external aid is provided to implement land reform.

The determinants of gains and losses after land property transfer

In order to simplify our understanding of the consequences of redistribution, we assume that those who are affected by land reform constitute two subgroups: gainers (or beneficiaries) and losers. The gains include expanded opportunities for the beneficiaries' income increase and the acquisition of command over food, self-respect and power to participate in village decision-making. Expected losses are incurred by landlords through the transfer of their property rights and from the partial or complete suppression of their rights to receive the market value of rent for the leased out land. Subject to each country's conditions, other losers are likely to include moneylenders, labour contractors and traders in irrigation water and farming machinery. There are other public gains, especially social stability in rural localities and the realization of national political stability.

Having said this, the *initial* scale of income transfer by state-led land reform in a given country is a cumulative result of a number of factors:

1 The extent of redistribution of privately owned land.
2 The scale and proportionate distribution of *publicly* owned land, in conjunction with the redistribution of *privately* owned land.
3 The gap or ratio between the average size of beneficiary holdings and the prescribed ceiling above which land area is distributed in the case of redistributive land reform, as well as the relative proportion of the sum of distributed land to total agricultural land.
4 The proportional size of the beneficiaries to the total number of agricultural households.
5 The extent of regulating sharecropping and tenancy (renting in and out) in the land lease market. The effect varies according to the proportionate reduction in rent below market value and the share of landowners. Are there 'shadow' rental values higher than the instituted levels?
6 Changes introduced to regulate water tenure rights and the terms of its use for irrigation.
7 Post-land property transfer arrangements for the marketing and processing of the new owners' products, individually or cooperatively.

But gains are not only materialistic. There are other gains in the psychological and social terrain. Examples are: (i) bases for popular participation, social uplift, self-respect and dignity; (ii) liberation or freedom from oppression and coercion; (iii) freedom from absolute dependency for survival upon the monopoly power of landlords, moneylenders and contractors of hired labour; and (iv) participation in political activities and in rural organizations, through the reform beneficiaries' own representatives in rural communities. Any observer of the pre-land reform oppression exercised by the cotton Pashas and royal estate managers of Egypt, the Zamindari of India, the tribal Sheikhs of Iraq and the owners of haciendas in the highlands or *altiplano* of Bolivia would appreciate these intangible social gains. An incident that occurred in the 1940s in the author's village in Egypt left a powerful impression on him during his childhood. A landlord hit a peasant *fellah* with a stick in front of his fellow villagers for what he considered a crime: the *fellah* did not dismount from his donkey when he passed the landlord who was sitting on his house porch. During the author's 3 years of work in Latin American countries in the 1960s, it became obvious to him that the *fellah* in Egypt was treated by his landlord no differently from the *inquilinos* and *campesinados* in Colombia, Peru and Paraguay.

Neoliberalism for voluntary land reform in a private market economy

Since the 1980s, there has been much discussion but little empirical analysis of the neoliberal foundations of market-based land policy as an integral instrument in the macroeconomic and global economic reforms, following the foreign debt crisis in developing countries. This land policy is presented as an alternative to genuine land reform, i.e. government-determined redistributive land reform. The writings on this revived faith in the global order of competitive market forces are suggestive of an amalgamation of the ideas of some moral philosophers (e.g. Robert Nozick) and a few Nobel Laureates in Economics (Buchanan, Friedman and Hayek). Despite their methodological differences, these distinguished social scientists share some ideas relevant to our inquiry. As I understand them, these ideas may be summarized as follows. In a dominant private property–free market economy, the state has a minimal role in economic activities or a limited interventionist role, irrespective of existing market failures. These failures exhibit an imperfect competitive market that is demonstrated by widespread unemployment and several forms of anarcho-capitalism, such as, for example, monopoly power, capital flight, concentration of land in a few hands, agricultural credit rationing, and so on. See Ritter (1980) for an analysis of anarcho-capitalism.

Conceptual foundations

To understand the nature of the amalgamated ideas behind the ideological shift towards the currently pursued market-orientated land policy, we examine briefly the ideas of one economist, Hayek, and those of one moral philosopher, Nozick, whose theories on entitlement and exploitation, related to land property, have been mentioned earlier. In his intellectual construction, Hayek (1978) stresses that a government is but one institution/organization. In his 'Great Open Society' (with capital letters), the state has two functions: (i) to enforce the rules set by individuals' free will and without coercion or central planning; and (ii) to perform services that the private sector cannot provide, particularly defence. In his view, any government intervention is detrimental to economic growth because only the market can coordinate economic activities for being self-equilibrating, whereas the government is a destabilizing force; a conception that arises mainly from the Austrian neoclassical model. His minimal state conception springs from vague assumptions formulated in a model that a social scientist, Forsyth (2002: 72), called 'Hayek's bizarre liberalism' (here, the state refers to an amalgamation of its constituencies including government, legislative and judicial systems).

With regard to the moral philosophers' ideas represented here by the theories of Robert Nozick (1974), they stress the moral and ethical foundations in a forceful, intellectual system of reasoning that lacks the economists' preference for measurable resource efficiency presented in quantitative models with a vague concern for equity issues.[8] As we shall soon find out, both groups of neoliberal social scientists have little or no obvious concern for the unfavourable distributional consequences of anarcho-capitalism/market mechanism free of government intervention referred to earlier. Nozick (1974: 113, 149) views the minimal state as the legitimate institution, having a monopoly for the protection of all individuals' rights and for establishing his theory of distributive justice, resulting from *voluntary* interpersonal exchange of property rights that is not deliberately designed by a central mechanism of the state (i.e. government) in bringing about a redistribution of landholdings. According to Nozick, land reform is, therefore, unnecessary. By considering voluntary exchange of property within the market mechanism and legal framework, his theory overlooks whatever consequences these principles might lead to in terms of extreme inequalities, prevalent undernourishment and persistent deprivation and absolute poverty of the propertyless, as well as the unfair gains and losses arising out of institutional monopoly explained earlier. His ideas on distributive justice also overlook whether the land grants made by former colonial rulers for their political convenience were ethically unjust although – according to his theory – they happened to be legally legitimate. In practice, it is important to trace the origin of present large holdings, that is how they were originally acquired prior to their legitimization.[9]

Land market in competitive national and international economies

Out of these simplified principles, and many of the neoclassical reasonings for competitive market economy – operating in a global free trade based on the theory of comparative advantages – came the foundations of market-based or market-assisted land reform and the aggressively pursued set of policies by what the 2002 Nobel Laureate in Economics, Joseph Stiglitz, calls 'The Washington Consensus'. It is the powerful trinity of the US government, the World Bank and the International Monetary Fund (IMF), to which I add the World Trade Organization (WTO), established in 1995 at Geneva. The pursued package of neoliberal policies includes macroeconomic structural adjustment, liberalized domestic and foreign trade, privatization, 'getting prices right' and managing budget deficits. This standard recipe is narrowly defined in the name of resource use efficiency and economic growth linked with free trade that has been applied as a standardized policy prescription to suit *all* developing countries.[10] Attempts to translate these principles into pilot projects in land policy will be examined later in this book (Chapter 7), and they have recently been well presented in a World Bank publication (Deininger 2003).

But, for the moment, we summarize their key elements. Different terminology is used to refer to market-based land policy: market-assisted land reform, negotiated land reform, market-friendly land reform, and so on. The main elements of this approach are: (i) direct land transactions between willing buyers from small farmers and sellers supported by borrowed foreign funds and financed by heavily subsidized credit in order to enable potential buyers to pay the going market prices; (ii) decentralization of programme implementation with the participation of non-governmental organizations (NGOs or civil society); (iii) no expropriation of private land property but, instead, a *voluntary* land transaction, using land title as collateral in the credit market, and (iv) land registration and titling. This land policy is associated with an obsession with the privatization of the indigenous customary land tenure that has, for centuries, been communally owned in many African and some Latin American countries. The privatization also includes lifting the restrictions on the sale of individualized communal land to outsiders.

Furthermore, the market-based land policy includes deforming the institutional arrangements of former state-mandated redistributive land reforms. Examples of deformation are: decollectivization of production cooperatives; dismantling state farms and preference for their sale on the open market; abrogating protective measures against the eviction of small tenants, requiring them to pay market-determined rental values.[11] The privatization of public domain lands and their sale to willing buyers – leaving inequality of private landownership distribution in the agricultural sector unchanged – with emphasis on environmental concerns has also emerged under the term 'The Other Land Reform' (Bromley 1989: 869–71).

This standard recipe for alleviating land inequality and poverty is assumed to operate by highly competitive landowners in competitive national and international economies which are considered necessary for the realization of sustained economic growth and enhanced exports. These conditions are made in a form of pro-market and anti-government intervention without a clear emphasis on their complementarity. What is missed or overlooked is the fact that, to function adequately, the market operation needs to be *enabled* by government actions in order to create investment and production incentives for land reform beneficiaries, and to provide traders with institutional facilities (e.g. customs regulations, licensing arrangements, pro-export rules, conducive land taxation, pro-small farmers' financial services and land registration facilities as well as the legal provisions for the protection of private property rights, and so on). In addition, the land reform beneficiaries, like other farmers, need government promotion of investments in physical land productivity and in human capabilities. These issues of interdependence and institutional arrangements for a complementarity between the workable market mechanism and government active intervention will be pointed out in Chapter 8, in the light of the findings of case studies.

Food security and the households' power of commanding their food needs

The preceding discussion on the different forms of land acquisition and the conceptual foundations of property rights transfer requires an understanding of the significant meaning of the resultant household food security, namely landholding-based freedom from fear and threat of food insecurity and hunger. From the dawn of human history, food security/command over food has been accounted as a major source of power and security. In *Power – A New Social Analysis*, Bertrand Russell (1938: 35) says, 'Power may be defined as the production of intended effects.' He also says, 'The impulse of submission, which is just as real and just as common as the impulse to command, has its roots in fear' (1938: 18).

In agrarian economies, a landless worker faces two types of fear and uncertainties in acquiring food. One is the unstable flow of income from hiring out his or her labour for temporary employment, and the other type of fear is his or her dependency on the power of grain traders in an imperfect market mechanism. In both cases, he or she is subordinated by fear and indignation. But, on the other hand, the peasant who owns or controls the use of a small piece of productive land has a higher degree of power and independence in acquiring most of the household's food from his or her holding. International relations seem to resemble this notion of power. A rich country donating food to a poor country commands a political power over, and imposes stringent regulations on, the latter, which submits and is subordinated.

Commanding food and productivity

Apart from being an investment in human capital, the nutritional improvement of a poor agricultural workforce (i.e. healthier labourers) raises its productiveness and in turn per capita real output. As indicated in Chapter 1, empirical evidence from several developing countries has established positive and statistically significant relations between better nutrition, labour productivity enhancement and faster economic growth (e.g. Arcand 2001; Taniguchi 2003). The researchers' conclusion is simple: malnourishment is costly in terms of slow economic growth. They add that, as nutritional improvement contributes to economic growth, policies providing command over food intake should be guaranteed for a long period of time (see Chapters 1 and 3 in this volume).

The relationships between nutritional improvement, economic growth and land access-based food commanding can be viewed as a product of three interrelated elements: security in acquiring the household's entire dietary needs; a higher degree of independence from the imperfect food market mechanism; and escaping the risk of malnutrition and, in turn, the risk of ill-health. The acquisition of food is *not* to be by means of its distribution by charitable organizations and government institutions. Rather, it is through an intensive mixing of the hitherto underutilized labour with legally secured access to productive farmland.[12] Certainly, *if* food for work and school feeding programmes, food subsidies and so forth are effectively executed in a stable manner and provided they are sustained over a long period of time, they can achieve good results. Nutritional benefits from these programmes may reach the rural poor, *not* necessarily the poorest, either directly or via temporary employment.

Why focus on absolute poverty in rural development?

Conventionally, rural development has a somewhat ambiguous and broader connotation than does land reform. As illustrated conceptually in Figure 2.1, it is a long-term process that can take generations to realize rather than decades. But the length of time could be considerably shortened if rural development started with a more equitable distribution of productive assets, notably land and education. By rural development, we mean the dynamic process of combined government action with the participation of low-income groups' own representatives to realize a rapid and sustained reduction in inequality and absolute poverty (we stress the decline in the *number* of individuals living in *absolute*, not relative, poverty). Absolute poverty refers to the number of people living below a cut-off level of an estimated income or consumption, satisfying the person's minimum necessities for a healthy life (notably food as the biological necessity for survival, as explained in Chapter 1).

A distinction between rural betterment and rural development

Poverty is reduced in proportionate and absolute terms. 'Proportionate poverty' is the *percentage* of the rural population (individuals or households) whose income/consumption falls below a poverty line identified and established by or for each country. In both cases, the reduction of poverty is measured by the same criteria (income or nutritional standard) between two or more points of time (t_1, t_2, . . . , t_n, where n denotes the number of years). If the percentage has declined but the number has increased, we call this change 'rural *betterment* in a transitional stage towards rural *development*'. During each phase, some poor individuals have become better off, rising above the poverty line. But for several reasons connected with the way inequality or poverty is generated, many others remain poor. Some, who were already above the poverty line, become poorer, eventually dropping below it. The balance and the exact composition of the poor and their ranking as poorer and poorest varies from one community to another within a country.

As an anti-poverty policy, land reform focuses on the poorest farmers (landless workers, small tenants, sharecroppers and owners of tiny plots). In developing countries, these subgroups seem to constitute the majority of the rural poor. For a complementary measurement of changes in poverty prevalence, the use of the Sen index is more meaningful, although its data requirements are demanding. This is because it combines the proportionate poverty prevalence with the ratio of the mean income or consumption of all the poor to the income or consumption poverty line. The combination also includes the measurement of inequality in income distribution among the poor (Sen 1981: Appendix C).

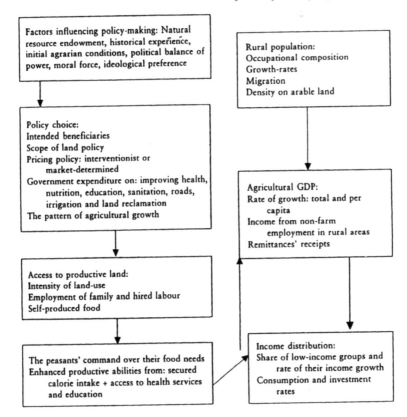

Figure 2.1 A conceptual illustration of the dynamics of land policy and the rural development process (adapted from Figure 3.2 in El-Ghonemy 1990a).

My proposed distinction between rural betterment and development may be clearer with a hypothetical illustration. Imagine two developing countries, X and Y, with equal rural populations and a high degree of concentration of land-ownership at an initial point of time t_1. Suppose also that the statisticians in both countries agree to use the same criteria for measuring absolute poverty at t_1 and t_2. Let us suppose further that the leadership in country X instituted and effectively implemented a substantial redistributive land reform to spearhead rural development immediately after t_1, whereas in country Y, the leadership did not because of a lack of political will. What follows is a hypothetical illustration of the relevant indicators, as shown in Table 2.1.

The variations over time between two imaginary cases is clear. Country X realized rural development, so defined; it reduced both proportionate poverty and number of the poor and, at the same time, it reduced, through land reform, the Gini index of land inequality. But despite its land reform, 20 per cent of country X's rural population are still poor because the eradication of absolute power is a long-term process and must address causes other than land concentration in

Table 2.1 A hypothetical index of land and income distribution for countries X and Y

	Poverty incidence		Inequality index (0–1)	
	Percentage of rural poor	No. of poor (millions)	Index of land distribution	Index of income distribution
Country X t_1	30	10	0.7	0.5
Country X t_2	20	8	0.3	0.2
Country Y t_1	40	10	0.7	0.5
Country Y t_2	33	13	0.8	0.5

a few hands (e.g. health, education, non-farm employment and labour-intensive technical change in agriculture). Country Y has attained rural betterment, but not development, that is the number of poor has increased despite the proportionate reduction. Inequality is still high, as the distribution of landownership has actually worsened through a process of land accumulation by the native and foreign landowners. In both countries, the battle against the twin problems of poverty and inequality is far from over.

Land reform dynamic linkages in the national economy

If we accept the thesis that changes in both the number of the poor and associated poverty characteristics are overall indicators of rural development, our distinction between 'betterment' and 'development' is not merely an academic one. By clarifying the meaning of rural development, we can further clarify its dynamic linkages with structural parameters in the national economy, as presented in Figure 2.1. The linkages suggest that rural poverty is a structural phenomenon, and that the variation in its prevalence over time is determined by forces outside the orbit of land reform. The links illustrated by arrows suggest that, for a large section of the agricultural labour force, increased productive capacity, earnings and effective demand for food, services and domestically manufactured non-agricultural commodities are determined not only by redistribution of land and power opportunities but also by other structural forces operating in the national economy.

Hence, there is feedback between the distribution of land/income/consumption/political power structure and rates of economic growth per head over time.[13] The order of magnitude of these changes depends on the scale of land reform and the rate of absorption of agricultural labour within rural areas and in the rest of the national economy. Fertility reduction, migration from rural to urban areas and the rates of growth in the agricultural labour force have a substantial effect on the differential rates of growth in per capita income between rural and urban sectors, and in the variations in the number of rural poor. As we shall explain in Chapter 8 from empirical evidence, population growth adds to the number of rural poor, who tend to have larger and younger (more children under the age of 5 years) families than do rich farmers.[14] Lacking social and economic security, these large, poor families need children to care for their elderly. With the notable decline in child and infant mortality, high fertility rates of rural populations delay a reduction in the number of rural poor.

Methodological problems in measurement

The foregoing sections suggest a number of difficulties in measuring the redistributive consequences of land reform. This is partly due to the multidimensional changes brought about within the dynamic forces of a changing economy and partly to the inevitable use of value judgements in the interpretation of social justice and injustice, liberty and coercion, participation and exclusion, and welfare considerations. Many of these difficulties are not only to be found in quantification, but also in finding consistent baseline data needed for temporal comparison before and after implementing land reform. In this section, we attempt an understanding of social justice as the basic structure of agrarian society, and discuss measurements of welfare and utility: changes in inequality in the distribution of land and income (or consumption); changes in motivation and participation; and issues in measuring the effect of income on food consumption and investment.

Social justice

Social justice, equality of opportunity and freedom of choice are phrases with interlinked meanings that are frequently used in both public statements and academic analysis. Besides, they have ethical and moral connotations. Yet, they are more complex than may appear. In our discussion of inequality of the distribution of agricultural land/income and the distribution of opportunities for land access, the investigation helps to illuminate some of the inequality and poverty issues and what is just or unjust and fair or unfair? How do the institutions in agrarian societies work to the disadvantages of the landless and the poor peasants?

In his theory of justice, Rawls (1972) holds the view that the benefits to the least advantaged in society justify reducing inequality or injustice (his difference principle). From the discussion on institutional monopoly, exploitation and inheritance, we have already seen the forms of inequality of opportunity and the violation of the principle of freedom of choice and the influence of such non-market forces as family background, enrichment by way of bequest and interfamily marriage. We shall discuss later in Chapters 5, 6, 7 and 8 how both the redistributive and market-based land reforms intend to remove or reduce the barriers to equality of opportunity for a landless worker's occupational mobility in the land tenure system to become a secure tenant and landowner. We need to keep in mind how these barriers manifest social injustice and how land reform brings in greater or better equality of opportunities to the hitherto disadvantaged rural groups.

Hence, the multidimensional character of inequality in the assessment of social justice that goes beyond the single dimension of a rural household's income/agricultural landholding inequality, judged later in the case studies by a single measure of the Gini index (from the Lorenz curve depicting the changes in the shares of landholders or owners by the size distribution of land). In fact, as argued by

Atkinson (1983) and Sen (2000), the assessment of social justice or injustice is a combination of several aspects of human welfare, including liberty/freedom of choice (personal and political rights), quality of life indicators (education, health, life expectancy or human capabilities which, together with income, have been developed by the UNDP in a human development index). Thus, the assessment of social justice or injustice in agrarian economies has been expanded beyond the narrow utilitarian examination, to which the discussion turns.

Welfare and utility

Without entering into conflicting arguments on utilitarianism and welfarism, let us attempt to understand it by way of considering a simple example. In agricultural society, X, the aggregate amount of farm land is fixed and the individual households are supposed to choose freely between three statuses of land tenure: to be a landless worker (L); to be an unsecured tenant (T); or to own land (O). Assume that:

1 the society prefers less inequality of income distribution than at present based on its ethical and value systems;
2 L has a lower level of welfare (income and social respect) than does T, who, in turn, has a lower level than O;
3 if L and T's incomes increase, so do their welfare and satisfaction from meeting their needs.

If the individual preferences are in the order of O is preferred to T, and T is preferred to L, the optimum state of welfare can accordingly be attained by redistributing the limited aggregate level of land (not necessarily to be equally distributed), provided that O is compensated for the loss incurred from transfer of land property rights, that all landless workers and tenants are included in the redistribution and that such transfer of property rights and income does not reduce total output.

Imagine three farmers, a landless worker who is illiterate, unhealthy, malnourished and has a large number of children; a tenant, who is illiterate (but healthy and not at risk of malnutrition) and owns two oxen and a plough; and an educated, healthy and well-fed landowner of 50 acres of productive land, 5 acres of which is rented out to the tenant farmer. Both the tenant and the labourer have different sizes of households. Hence, the three have different economic and non-economic characteristics. Each has a unique utility function in response to their different tastes, abilities and incomes. As is to be expected, there is a conflict of interests among the three people, namely with reference to the worker's terms of his labour utilization, the tenant's leasing-in arrangements and the landowner's rational entrepreneurial motives for making the maximum profit out of his transactions with the two peasants.

In response to the preference of the society indicated earlier, and in order to attain a high level of potential total welfare, the society leadership decides to take

10 acres from the landowner out of his or her total area of 50 acres and transfer its ownership equally to the two peasants who do not have their plots free of charge. The redistributive agency of the society serves as a broker and pays the affected landowner the price of the 10 acres on long-term instalments. Being a shrewd entrepreneur, he works harder on his remaining 40 acres to raise its output. The two jubilant peasants use their formerly unutilized family labour and newly acquired access to complementary inputs and also raise the productivity of their holdings. They now enjoy a higher income and social status than before. The sum of the three individuals' benefits is likely to be higher than before the land redistribution. Other things being equal, if the benefits derived from this limited example are extended to a larger number of households, the sum of the individual households' benefits is likely to be higher. The welfare of those who are now better off is higher than the dissatisfaction experienced by the ex-landlords, and total output is higher than before.

This oversimplified reasoning is based on a number of assumptions about society X, and three individuals, and upon a moral judgement. Claims have been made without quantification of different interpersonal utilities of those whose social welfare has uniformly increased, and their needs met. Nor did we offer a calculus of the decline in the utility of those whose social status and property were diminished. Conventional utilitarianism and utility-based welfarism cannot help us either. As we understand this system of thought, it is concerned only with the sum of individual utilities (social welfare) and not its equitable distribution among individuals.[15] The theory assumes that all losers (landlords) are homogeneous and all gainers (peasants) are also homogeneous. It considers that members of each subgroup have a similar utility function, and that their interests are identical. It ignores their differences in initial inequalities in the ownership of assets and in human quality. Equally, it ignores non-utility considerations of liberty, power, self-respect, exploitation and coercive relationships, which were stressed earlier as notions of special concern to land reform.

Land and income distribution measurement

The measurement difficulties of welfare and utility do not exist in measuring changes in the distribution of landownership and operational holdings (leased and sharecropped). Agricultural censuses provide data on the number and area of holdings by class size and tenure status (owners, tenants, labourers, managers of large farms with absentee owners). The change in the degree of concentration may be measured statistically by the Gini coefficient and ratio of the average (mean or median) size of holdings in the highest to the lowest size category of holdings. Land*ownership* distribution is usually more unequal than the distribution of *holdings* because some of the owned land is rented out to be operated by several tenants.

But land is only a means for securing the household income or consumption and commanding its food needs. Identifying inequality in income distribution as well as differences in the composition of income and the size of sources of non-land income is therefore necessary. To analyse inequality and compare incomes, we need to recognize differences in the composition of income and the size of households. Per capita income is, therefore, preferable to total household income. Current gross income is preferable to current cash income because of the need to include remittances and self-grown food. (It usually does not, however, include government-subsidized services and other benefits from public expenditure.) Total consumption is preferable to income because it indicates the components (food, clothes, transport, fuel, education, health, etc.), the shares of which are important indicators of changes in living standard. From carefully conducted household expenditure surveys, we have data on 'what' and 'how' much individuals consume.

Gini and Theil indices of inequality in income distribution are powerful tools in judging changes in inequality in the income (or consumption) shares of groups and inequality among each group.[16] Provided that such data on income distribution are available and comparable (which is not easy), we can judge the extent of changes in the consumption pattern and in inequality before and after land reform. But how long after? The reduction in the concentration of *landownership* is easily measurable soon after the completion of the land property transfer. But to measure the decline in inequality in the distribution of income or expenditure and changes in undernutrition and power structure are more problematical and require observations over a longer period of time. Complete land redistribution, enforcing tenancy regulations and arranging for irrigation, drainage, supply of complementary inputs and marketing of beneficiaries' produce takes, on average, 5–8 years. It appears that a total period of 15 years should provide enough time for: (i) the stabilization of the relative economic position of losers and gainers from land reform; (ii) the beneficiaries to experience their new responsibilities in production, marketing, capital formation and participation in the development of their communities; and (iii) the manifestation of improved abilities as a consequence of new motivations, better nutritional standards and developed skills.

Distributional consequences: consumption and saving

The economic behaviour of the land reform beneficiaries as consumers and potential savers depends upon the extent of their pre-land reform deprivation. The beneficiaries who were poor, underemployed and malnourished are likely to eat more and better. With the expected rise in their income, the allocation for the two necessities (food and fuel) is likely to remain proportionately constant during the initial phase of the reform (the food share is likely to remain at about 70 per cent of total household expenditure because of the expected high elasticity of demand for food – around 0.8).

We know from the ideas behind Ernst Engels' law (1857) and Keynes' marginal propensity to consume (1936) that a rise in the beneficiaries' income leads to a steady fall in their income elasticity of demand for food and a reciprocal rise in the demand for non-food items.[17] As noted earlier, this relationship is not expected to hold at an early stage of land reform. This has been substantiated by empirical research since the original work of Clark and Haswell (1964) on the consumption behaviour of the poor.

Given that 70–75 per cent of the beneficiaries' total income is spent on food, the balance of their total outlay will be allocated to the purchase of non-food items (clothes, furniture, bicycles, etc.) which are normally *domestic* products. There is no reason to believe that they would sacrifice their own present consumption for a consequent investment to increase future income. Increased consumption is a logical prerequisite to reducing the poverty endured before land reform. However, at a later stage, when their needs are met, the balance for non-food expenditure and savings for productive investment is expected to rise. The higher the aggregate demand (purchase) for locally produced non-food items, the greater is the employment of labour in the corresponding industries and services in the domestic economy.[18] This may be illustrated by the hypothetical example in Table 2.2.

This shows that the land reform beneficiaries are contributing to the productive capacity of their national economy via their consumption of non-food items and their saving to add new non-land assets or investment goods (e.g. purchase of livestock, irrigation pumps or shares in their cooperatives). The order of magnitude of these income effects depends primarily on the type and scale of land reform.

For the affected landlords, the redistributive consequences depend on many factors: Is land reform complete or partial?[19] How much was their pre-reform saving capacity? Are they compensated or not for the expropriated land and other farming assets? Is compensation an outright payment in cash or in government bonds? What is the rate of inflation in the post-reform period? In countries where they are compensated, it would be highly unlikely for them to invest extensively in agriculture for fear of future expropriation. If compensation was in cash, it would be sensible for the government to prevent its flight abroad and ensure its injection into the domestic economy through a number of outlets: savings in financial institutions and investment in industry, trade, services and housing construction in urban areas. Being experienced entrepreneurs, the affected landlords are likely to increase productivity from the retained advantageously situated and higher quality land compared with that of the beneficiaries after expropriation.

Summary

This chapter has examined the principles behind the current debate surrounding market liberalization and the pros and cons of government intervention in

Table 2.2 Hypothetical example of the consumption behaviour of the poor

	Before land reform	After land reform	
		Period I	Period II
Value of output/income	100	130	150
Consumption – food	75	95	100
Consumption – non-food	25	30	40
Savings (investment)	0	5	10

reforming the land tenure system. Several widely used terms/concepts, with different interpretations in relation to land property rights and their adjustment, have been specified. In particular, the post-1980s shift away from state-administered land reform and towards market-based land policy has been discussed and its possible foundations explored.

To understand the arguments of the supporters and opponents of both sides of the current policy debate, the chapter has explained such important concepts as institutional monopoly, barriers to entry into the agricultural land and credit markets, non-market land property transfer, commanding household food needs through secure land access, sustaining land reform beneficiaries' gains and minimizing losses, and the intertemporal and intercountry comparability problems of changes in poverty, malnourishment and the inequality of land distribution.

It is argued that the post-1980 increased prevalence of poverty, undernourishment, land concentration in a few hands and landlessness requires taking the debate beyond its narrow context of pro-market-based land reform or anti-government redistributive land reform. Institutional arrangements for their complementarity within an anti-poverty rural development strategy are suggested, and will be empirically verified by the findings of 16 case studies. These elements constitute the analytical frame for the rest of the book, keeping the central issue as not merely freeing the market for economic growth but how best to rapidly remove institutional monopoly and reduce absolute poverty, malnourishment and landlessness by way of secure land access. The interdependence between the institutions of market and government and how their complementarity works is examined further in Chapter 8.

Suggested reading

Atkinson, A.B. (1975) *The Economics of Inequality*, Oxford: Oxford University Press.
—— (1983) *Social Justice and Public Policy*, Cambridge, MA: MIT Press.
Baranzini, Mauro (1991) *A Theory of Wealth Distribution and Accumulation*, Oxford: Clarendon Press.
Barry, Norman (2002) 'The New Liberalism' in Smith, G. (ed.) *Liberalism*, Vol. II, London: Routledge.
Bromley, Daniel (1989) 'The Other Land Reform', *World Development*, 17(6).
Carter, M. (2000) 'Old Questions and New Realities: Land in Post-Liberal Economics' in Zoomers, A. and van de Haar, G. (eds) *Current Land Policy in Latin America*.
El-Ghonemy, M. Riad (1990) *The Political Economy of Rural Poverty*, London: Routledge.
—— (1999) *The Political Economy of Market-Based Land Reform*, Geneva: UNRISD.
Nozick, Robert (1974) *Anarchy, State and Utopia*, Oxford: Basil Blackwell.
Roemer, John (1996) *Theories of Distributive Justice*, Cambridge, MA: Harvard University Press.
Sen, Amartya (2000) 'Social Justice and the Distribution of Income' in Atkinson, A. and Bourguignon, F. (eds) *Handbook of Income Distribution*, Vol. 1, Amsterdam: North Holland.
Stiglitz, Joseph (2001) 'Redefining the Role of the State', Tokyo: MITI, published

as Chapter 3 in Chang, H. (ed.) *Stiglitz: The Rebel Within*, London: Anthem World Economics.

Taniguchi, K. and Wang, X. (eds) (2003) *Nutrition Intake and Economic Growth*, Rome: FAO.

Wade, R. (1990) *Governing the Market*, Princeton, NJ: Princeton University Press.

Waldron, Jeremy (1988) *The Right to Private Property*, Oxford: Clarendon Press.

Warriner, Doreen (1969) *Land Reform in Principle and Practice*, Oxford: Clarendon Press.

3   The agricultural dimension of poverty

Understanding the causal process of rural poverty requires an examination of its agricultural dimension, which comprises several factors: the supply of arable land and irrigation water; food production relative to human nutritional needs; employment opportunities for the agricultural labour force; investment in human and land productivity; the seasonality of agricultural activities; the vulnerability to climate change; and the fluctuating terms of trade. In addition, and importantly, the agricultural dimension includes its institutional and production organization with emphasis on land property rights and tenure arrangements, and the scale of landlessness in relation to the growth rates of the agricultural labour force and arable land supply expansion. We begin with food, the biological necessity for human survival. The intake of food that satisfies the minimum requirements is a major yardstick in the measurement of poverty prevalence, as specified in Chapter 1.

Access to land and commanding food

In 2005 and 2006, we have witnessed the severity of both undernutrition and ill-health combined with the collapse of food consumption, starvation, extreme food shortage and depletion of human capital leading to death in Niger, Darfur (Sudan) and Somalia, owing to food production failure brought forth by a combination of locust invasion, prolonged drought and armed tribal disputes. Likewise, researchers of recorded famines in the twentieth century have found unequivocal evidence that most of the victims have been the rural poor in general and landless wage workers in particular. The 1998 Nobel Laureate in Economics, Amartya Sen, has shown that, even in conditions of adequate food supplies, in 1981 for example, starvation can occur among those groups who lack the resources to buy food. His analysis of four famines – in Bengal in 1943, Ethiopia in 1972–74, the African Sahel in 1968–73 and Bangladesh in 1974 – demonstrated that only in the Sahel did a decline in actual food availability contribute to the disaster. In the other three famine situations, food availability per person was no worse or even better than in non-famine years. Sen found that the victims had one factor in common: the ratio of the price of food grains to their wage rates, or to the other goods they were producing or services they were offering, moved sharply against them, due to grain traders' inflationary pressure on food grains.

Famine, land access, malnourishment and poverty linkages

During famine, several factors were at work. First, food price hikes resulting from a shortfall in supply forced a larger number of the rural poor, including landless workers, small tenants and pastoralists, to seek supplementary employment. Consequently, the workforce expanded at a time when demand for agricultural labour was low, with the result that wages declined or were stagnant. A second factor was that, as people sold off their livestock, tools, jewellery and even land in order to pay for food, the market value of the assets tended to fall. Finally, in famine situations, price ratios of other foods and services tended to fall sharply against grain, while the demand for services also slumped. Thus, the victims were those with few resources who were in low-income wage employment. In Bengal, they were 'fishermen, transport workers, paddy huskers, landless agricultural labourers, craftsmen and non-agricultural labourers'. In Ethiopia, the occupational categories of the destitute included pastoralists, evicted farm servants and dependants of rural labourers, tenant cultivators and small landowning cultivators. In Bangladesh, Alamgir (1980) found that, during 1973–74, small tenants and landless workers were particularly vulnerable to famine conditions.

Famine can generate or increase permanent poverty because its victims lose assets and future income through livestock deaths, the distress sale of assets and debt. The result is usually an increase in the numbers of people living below the estimated country-specific poverty line. In a single year, the proportion of people in 'extreme poverty' in Bangladesh rose from 25 per cent to 42 per cent between 1973 and 1974, while landlessness increased by about 3 per cent. The position of the rural elite landlords and grain traders was strengthened after the famine (Sen 1981; Alamgir 1980).

This last point of how the market power of grain traders increases peasants' deaths during famine is the main thrust of Martin Ravallion's investigation (1987). On the basis of his careful investigation of two famines, in south India in 1977 and in Bangladesh in 1974, Ravallion identified high food grain prices as an important cause of starvation and deaths (1987: Figures 2.6 and 2.7). In examining the relationship between *aggregate* food availability and individual survival chance, he finds that the high food grain prices combined with a fall in employment determine the agricultural workers' survival chance, irrespective of high food grain stocks in the country. His econometric analysis indicates that, during the Bangladesh famine (1974), a 10 per cent drop in employment resulted in a 21 per cent fall in the *individual's* consumption of rice (Ravallion 1987: Table 1.2). Thus, peasants who have no accessible opportunities to hold productive land and consequently fail to establish command over their food needs are the most likely candidates for malnourishment, high risk of hunger and severe ill-health leading to death.

Climatic conditions cannot be held entirely responsible. In his careful study of the 1984–85 famine in Darfur (Sudan), De Waal (1989) ranked the drought as the most obvious and 'immediate cause' of famine, whereas the 97 per cent of local people surveyed were of the opinion that 'the drought was a suffering sent from God as a punishment' (ibid.: 78–9). Other observers viewed the repeated drought in terms of climate change: declining overall amount of rain, changing pattern of rainfall, sandy winds becoming more frequent, and so on. According to De Waal, the suffering of the people of Darfur manifested a trinity of hunger, destitution and death that totalled nearly 100,000 as an excess of expected average crude deaths. Furthermore, child mortality below the age of 5 years increased in 1 year from 63 per 1,000 to 282 per 1,000, i.e. a sixfold increase in deaths in a *single* year (ibid.: 178).

Vulnerability of the landless to undernourishment in normal times

The empirical evidence of the several famines cited here suggests that those who starved to death have been landless agricultural workers, small tenants and pastoral nomads. They also suggest that, in normal times of relative plenty, landless agricultural workers not receiving the wage equivalent in grain are most vulnerable to suffering from undernourishment and at high risk of ill-health under the following conditions:

1   They have to rely on the labour market to exchange their labour for wages and on the grain market to exchange their wages for the purchase of household food requirements (Sen 1981).
2   They are vulnerable to displacement by capital-intensive technology in production (e.g. mechanization).
3   The seasonality in food production leads to involuntary unemployment and low calorie intake *before* harvest time when food reserves are low and prices are high (Longhurst 1983).
4   The type of labour in farming is physically demanding. These demands on their physical energy may be in addition to the calorie loss incurred by walking the long distances to work. This is also true for the distance travelled by many rural women to fetch water and by many children to school. These three types of compulsory physical efforts in agriculture require calories from an already deficient supply. (This is in sharp contrast to the urban elite, who try hard to lose their excessive calorie intake by jogging or playing squash.)
5   In the absence of remittance receipts and social security schemes, the subsistence-level wages of the landless peasants leave no balance for social needs. This throws them into debt. The nutritional effect of such distress is more serious in cold climates where the demand for clothing competes with the demand for food (El-Ghonemy 1990a, 1993).

Key structural problems in agriculture

In this section, four important agricultural structural issues are illuminated: neglect of food production in general and food grains on small farms in particular; distorted investment in agriculture; the deficient institutional framework of agriculture; and an existing trend of fast-growing demand for – and declining supply of – cultivable land in most developing countries. In this inquiry, aggregative data from developing countries are used to judge these intertemporal changes.

Table 3.1 Performance of agriculture in developing countries, 1965–2003

			1965–80 (%)	1980–85 (%)	1990–2003 (%)
*Population: average annual growth rate (%)					
Low-income economies			2.3	1.9	2.0
Middle-income economies			2.4	2.3	1.1
†Agricultural GDP growth rate (%)					
Low-income economies					
Total			2.7	6.0	3.0
Excluding China and India			2.0	1.9	
Per capita			0.3	1.1	1.0
‡Arable land (hectares) per capita agricultural population	1974–76	0.36		0.34	0.21
Annual rate of growth (% of above)	1968–80	–0.25		0.56	–0.03

Sources
*World Development Report 1987, Development Indicators, and 1990–2003 calculated from *World Development Indicators* 2005 and FAO *Production Yearbook* Vol. 57, 2003.
†*World Development Report*, 1982: 41 and *World Development Indicators* 2005.
‡FAO, *Production Yearbook*, 1985, in Table 2, FAO (C 87.89) period, 1971–73 and 1980–82 from FAO, The State of Food and Agriculture (SOFA), 1985, Tables 1–6.

Notes
Arable or cultivable land is the amount of land available for cultivation irrespective of its intensive use. Agricultural population is those adults engaged in agriculture as their main occupation and their dependants.

Some of these aggregate indicators are presented in Table 3.1 to provide the reader with an overall picture. Total population annual growth is still high at 2 per cent, and the economically engaged population in agricultural activities represents 45 per cent of the total workforce. If their dependants are added, they become the majority of the total population. Average annual growth rates of agricultural gross domestic product (GDP), both total and per capita, fell between 1980 and 2003. The same downward tendency but at a faster rate is observed in arable land, both total and per head of agricultural population, from –0.25 per cent in 1960–80 to –0.03 per cent on average in 1990–2003. Later in this chapter, we shall examine the likely implications of this alarming trend for meeting the growing demand for land access. But let us begin with the critical question of falling food production.

Declining food production, 1960–2003

As we have repeatedly stated, food is different from other agricultural commodities, which may explain why so many developing countries insist that food security/food self-sufficiency in relation to population growth is set as a major policy objective, irrespective of the country's global product competitiveness. Since Thomas Malthus' and David Ricardo's hypotheses in the late eighteenth and early nineteenth centuries, this question has been one of the long-debated issues in development in a world of speedy climate change, increasing scarcity and rapid technological advance.

Despite the ideals of food self-sufficiency and food security, the empirical evidence in 2002 and since 1960 is alarming. During the 25-year period 1960–85, of 128 developing countries, only 30 per cent or 38 countries experienced food production per head of total population (food productivity) that continuously kept up with population growth over the entire period. The remaining 90 countries include 27 whose percentage of undernourished people increased between 1969–71 and 1979–81 (FAO 1985a: Table 2.2 and Appendix 1) and 24 countries in which over 50 per cent of the rural population live in absolute poverty. This distressing outcome of the neglect of food production continued into 2002. Between average 1989–91 and average 2000–02, of the total of 129 developing countries, 66 experienced a decline in food productivity, while 63 countries (49 per cent) were able to keep up with population growth. What is worrying indeed is that over half the countries with falling food productivity were in Africa (FAO 2003a: Table 9).

Breaking the aggregate data on food production into regions, Africa experienced the most serious decline because it was continuous throughout the long period 1960–2002. It is remarkable that, in Asia, with the most populated countries in the world, food productivity improved substantially. Furthermore, taking a closer look at the countries whose economies were centrally planned/socialist and whose agrarian system is more egalitarian, we find from a sample of eight countries given in Table 3.2 that total and per head food production during 1960–2003 was higher than in the rest of the developing countries. We should note that the table includes two private property market economies that have had a redistributive land policy since the 1950s and 1960s (South Korea and Libya respectively). I shall return to this question of equity growth relationship after examining the trend in the inequality of land distribution from 1950 to 2004.

Elsewhere (El-Ghonemy 1990a: Table 1.4), I have studied the situation in 19 African countries with available estimates of poverty levels (50 per cent and above) and having 68 per cent of their total labour force employed in agriculture and 80 per cent of the population living in rural areas. The study shows that, out of the 19 African countries, 74 per cent or 14 countries had negative growth rates of food production per head in 1970–84 and, in three others, the margin was between 0.1 and 0.4 during the same period. The study concludes, 'Certainly there is justification for growing national and international concerns with many developing countries' failure to match their population's rising demand for food' (ibid.: 27).

Neglect of food grain production in small farms

The results of the agricultural censuses since 1950 reveal two features with respect to land use. First, they show that countries characterized by a higher percentage

Table 3.2 Food production in eight developing countries with egalitarian agrarian systems, 1960–2003

| | Annual rate of growth | | | | | | |
| | Total food production | | | Per head | | | |
	1960–70	1971–80	1981–90	1960–70	1971–80	1981–85	2000–03
Albania	3.2	4.6	2.6	0.4	2.1	0.4	2.5
China	5.8	3.1	6.0	3.4	1.3	4.7	2.5
Cuba	4.3	3.0	2.1	2.2	1.8	1.5	6.0
Laos	6.0	2.6	6.5	3.3	0.7	4.2	0.5
Libya	7.0	5.3	11.9	2.9	1.2	7.7	1.2
North Korea		5.7	4.2		3.0	1.7	3.5
South Korea	4.0	4.8	3.7	1.5	3.0	2.1	1.0
Vietnam	1.3	3.4	5.1	-0.8	1.0	3.1	6.3
Average of all developing countries		2.8	2.5				-0.4

Sources: Calculated from FAO, *Production Yearbook* and *World Development Indicators*, several years (see sources of Table 3.1).

of large landholdings tend to have little cropland and more land under pastures or left temporarily fallow. In South America, for example, this fallow land represents nearly 65 per cent of the total landholdings area, whereas 20–45 per cent of the total population depending on agriculture for livelihood are crowded in the small cropped area, and many of them are caught in the poverty trap. In contrast, the percentage of cropland to total area of holdings was 92 per cent in Asia and the Middle East, two land-scarce regions where the area cultivated by temporary crops, mostly food crops, was 86 per cent. Second, it is smallholders with less than 1 hectare who are the primary growers of food grains. For instance, a detailed examination of the 1970 and 1990 agricultural censuses shows that this group of landholders produced 74 per cent of total harvested wheat, 68 per cent of rice and 60 per cent of maize.

This high intensity of food cropping in small farms has been made possible despite several existing institutional constraints in the supply of credit and technical services; once removed, their productivity was realized, as, for example, in the cases of most effectively implemented land reform programmes. Evidence shows that, among small farmers, it is women who are particularly constrained in many developing countries, where they are denied access to institutional credit and technical services as widows operating landholdings and, in many cases, they are agricultural household heads during their husbands' migration. Before 2002, agricultural censuses did not disaggregate farmholders by sex, and sex differentiation of access to land and credit has usually been ignored in data collection and tabulation processes. Available household surveys estimate female heads of household at 15–20 per cent, with 50–60 per cent of working time on food-producing farms worked by women. These percentages are higher among smallholdings, especially in areas with widespread migration of male heads of the households (Safilios-Rothschild 1982, 1983; FAO 1999). Yet, women as significant producers of food crops, livestock and dairy products are virtually invisible to policy-makers and bureaucrats in most developing countries. In the next chapter, I shall return to these issues of investment, productivity and cropping intensity by size of agricultural landholdings irrespective of gender.

Distorted investment in agriculture

From the results of countries' experiences over the last four decades, it seems that the neglect of female food entrepreneurs and small grain producers is a part of the general investment inadequacy in agriculture that is detrimental to agricultural populations' productivity (low average productivity), which contributes to persistent poverty via low income and low purchasing power, particularly in rainfed agriculture. The extent of inadequacy that I have used in my 48 developing countries' studies of 1984 and 1990, and that of Egypt (1976–2003), is estimated by relating the share of agriculture in gross fixed capital formation (GFCF)[1] to its share in national income (GDP) and total labour force, and by the share of agriculture in total public expenditure as an indication of the extent of government commitment to agriculture (i.e. expenditures on irrigation, land reclamation, soil conservation, drainage, land title registration and road construction that is essential for transporting the products to market centres, and so on). The former measurement of GFCF is preferable to the latter because it includes capital formation from both private and public subsectors and is comparable to that calculated in the *United Nations National Accounts Statistics*. For example, I found that Egypt's average GFCF in 1980–2000 was 7 per cent and that it should have received 21 per cent.[2]

About 250 years ago, Francois Quesnay, the leader of the Physiocrats, addressed this subject saying:

> Everything that is disadvantageous to agriculture is prejudicial to the state and the nation, and everything that favours agriculture is profitable to the state and the nation [. . .] It is agriculture which furnishes the material of industry and commerce and which pays both.
>
> Alexander Gray (1931: 102)

This argument for the accentuation of France's agriculture and its linkages to other sectors of the economy in the early 1700s is still very relevant today to many developing countries, particularly those that have a low income, a food deficit and persistent rural poverty. In economic terms, there is no justification for underestimating agriculture or for not recognizing its capital needs in order to accelerate agricultural growth rates and, in turn, the productivity of the agricultural workforce. The principle of this long-established relationship was formulated by the Harrod–Domar model in the 1940s and modified by Robert Solow, the 1987 Nobel Laureate in Economics. The realities of declining international aid, particularly to agriculture and rural development, contrary to the donors' proclaimed commitments, have been presented in Chapter 1.

Growing demand for, and declining supply of, land: shrinking area per working person in agriculture

This section examines what happened to the supply of agricultural land relative to the growing numbers of people working in agriculture between 1970 and 2003 (with a projection for the year 2010). This trend is significant to the understanding of land access prospects, including those for the land market approach. Table 3.3 presents data for a sample of 13 developing countries for which data are available for this long period. Because comparable data on land prices (rental and purchase) are lacking, the ratio of arable land (in hectares) per working person in agriculture is used as an approximation of the supply of and the demand for land. In this context, it is recognized that the aggregation of land conceals wide variations in quality, cropping intensity and capitalized value between and within countries. In addition, the use of averages obscures two factors: rapid urbanization, which takes scarce cropland out of agriculture for non-agricultural purposes; and the increasing costs of land reclamation for the establishment of new irrigated land settlement schemes, which are particularly costly in North Africa and the Middle East because of aridity, soil texture and water scarcity.[3] We have already seen

Table 3.3 Changes in arable land and pressure of agricultural workforce on land in 13 developing countries, 1970–2002, and projection for 2010

Countries in alphabetical order	Arable land					Annual growth of agricultural workforce (%) (3)			Ratio of used arable land area to agricultural workforce (hectares/person) (4)			
	Annual growth of used arable land (%) (1)				Area of balance for future crop production as a percentage of total land with crop production potential (2)							
	1970–80	1981–90	1991–96	1995–2002		1980–90	1990–2000	2000–10	1970	1980	1996	2002
Algeria	1.2	0.3	1.2	0.5	25	0.9	1.2	0.5	4.9	5.9	3.9	2.9
Brazil	3.7	1.7	3.0	0.6	85	-0.3	-0.7	-1.1	2.4	3.5	3.7	2.3
Colombia	0.3	0.4	0.3	-0.1	88	0.6	-0.2	-0.9	2.1	1.9	1.5	0.6
Egypt	0.6	0.0	4.0	0.0	3	0.9	1.2	0.5	0.6	0.5	0.4	0.3
Ethiopia	0.3	0.0	-0.2	0.0	57	1.2	1.0	1.2	1.1	1.0	0.7	0.4
Kenya	1.1	0.8	0.0	0.0	50	2.7	2.8	2.6	0.5	0.4	0.4	0.4
Malawi	0.8	0.3	0.2	0.1	59	1.6	1.4	1.2	0.6	0.6	0.4	0.4
Mauritania	-3.9	0.5	0.2	0.0	65	1.8	2.0	2.3	0.8	0.6	0.4	0.8

Morocco	2.5	1.6	0.5	−0.5	50	0.9	0.5	0.1	3.2	3.1	2.2	1.9
Philippines	0.9	0.2	0.1	0.0	42	1.5	1.4	1.1	0.9	0.8	0.8	0.5
South Africa	1.0	0.8	0.7	−0.1	–	0.8	−0.6	−0.9	4.8	8.1	7.1	8.8
Sudan	0.7	0.4	0.1	2.4	82	1.3	0.6	0.9	3.2	2.9	1.8	2.0
Uganda	1.5	2.1	0.2	0.0	58	2.2	2.3	2.2	1.2	1.1	0.8	0.5

Sources: Column (1) is calculated from FAO (1993a, 1996b); column (2) is based on FAO (1993b: Table A.5); column (3) is taken from FAO (1993b: Table A.1); column (4) is calculated from same sources as column (1) except 1995–2002 is calculated from Alexandratos (1988), FAO (1993b, 2003b) and from the World Bank (2005). In these sources, the agricultural workforce is termed 'the economically active population in agriculture (farming, animal husbandry, forestry and fishing)'.

Notes

'Arable' or cultivable land is land cultivated with temporary and permanent crops, and land under temporary fallow: it does not include forest and permanent pasture lands. The ILO *Yearbook of Statistics* defines the economically active population as those persons who provide labour either employed or unemployed. Labour force or workforce is those currently or usually active for a period longer than one day/one week. 0.0 per cent annual growth means either no change or a change of less than 0.01 per cent.

Potential land or 'balance' is land of varying quality with potential for growing crops; it is a rough estimate and comprises land in actual crop production use (rainfed and irrigated) and land that could be cultivated in future.

from the data given in Table 3.1 the fast-shrinking area of arable land per person of agricultural workforce in developing countries from an average of 0.36 hectares in 1974–76 to 0.21 hectares, on average, in 1990–2003.

Let us now have a closer look at the data presented in Table 3.3 because they are important to the rural poor's opportunities for access to land explored in Chapter 8. With the exception of land-abundant Brazil, Colombia and Sudan, and mineral-rich South Africa, there is a very low ratio of actually used arable land to agricultural workforce, particularly in Egypt, Kenya and Malawi. There is also a general downward trend in this ratio, notably during the last two decades between 1980 and 2002 in most countries, especially Egypt and Morocco. One possible explanation for this decline is the slow growth or stagnation in cropland expansion in many countries while the population/agricultural workforce is growing at fast rates, especially in African countries. Moreover, post-1980 structural adjustment, price liberalization and fiscal reforms have required a minimal role for the state in economic activities and also heavy cuts in public expenditure, including land-augmenting irrigation and the expansion of land settlement schemes. In my examination of FAO data on irrigation expansion between 1989 and 2002, I found that, of the total 140 developing countries for which data are available, nearly 90 per cent manifested an alarming stagnation and/or decline in irrigation expansion (i.e. total irrigated area), especially in the arid and semi-arid regions of North Africa and the Middle East. I also found that about 53 per cent of the Middle Eastern population live in areas with less than the acceptable minimum level of 1,000 cubic metres of water availability per person per year and, if present rates of use continue, the average is expected to be halved by the year 2025 (FAO 2003a: Table 2; El-Ghonemy 1998: 55).

What is of great concern is the fact that, in poor countries whose financial capacity to invest in land supply expansion is very limited (Ethiopia, Kenya, Malawi, Mauritania, Sudan and Uganda), there are relatively large, uncropped areas that are potentially suitable for crop production subject to the availability of capital (see column 2 of Table 3.3). The large potential area in Colombia and Brazil is a manifestation of the widespread practice of absentee landownership, with severely underutilized large estates. According to the results of the 1990 Agricultural Census of Colombia, the number of landholdings in the category of over 200 hectares accounts for only 5 per cent of the total, but their area represents 54 per cent of the total (see Chapter 7 for potential land access to the rural poor in Colombia).

In the light of remarkable technical advances for bringing land, water and even human fertility under the control of applied science, it is incredible that climate and population growth are still identified by politicians as the lone culprits.

The deficient institutional framework of agriculture

Leaving for the moment the impact of the recent cycles of economic reforms and their impact on public expenditure on irrigation and agricultural commodity pricing, the intricate obstacles set by the institutional framework of production, exchange and distribution are examined in this section. Many developing countries have adverse institutional influences that take different forms. They include the creation of a state of underutilization of labour, land and technology in agriculture and incentive restraints of production and inducement to invest that are needed to raise the demand for labour, and to realize their potential productivity benefits translated into high demand for goods and services produced domestically outside agriculture. On the other hand, neglect or failing of the constituents of this chain contributes to rural poverty via the negative effects of undernutrition in the agricultural workforce leading to very low levels of both productivity and individual income/consumption of a large section of the existing workforce (i.e. their low purchasing power).

Increasing land concentration and polarization[4]

In this section, we are primarily concerned with the size distribution of landholding units in which the scale and use of the means of production are at work. Leaving the exploration of how monopoly powers and land concentration have come about to country case studies in Chapters 5, 6 and 7, we consider here the aggregate features of the size distribution of landholdings (owned, rented, sharecropped or a combination of these tenure arrangements). Data on holdings are given in country agricultural censuses that are internationally designed and analysed by the FAO of the United Nations and uniformly report area and size distribution. The available results for 23 countries that completed at least two censuses including those during 1978–84 and 2000–04 were compiled by the present author and are presented in Table 3.4.[5] They suggest a few generalities: first, that there is rising inequality in the distribution of landholdings; second, that there is an increasing polarization in agrarian structure.

Furthermore, another measurement of increased inequality shown by the results of agricultural censuses is the share of small landholdings (below 5 hectares), which has increased, whereas the corresponding areas did not. Contrarily, the actual size of large holdings grew over the same period. In fact, my inspection of the detailed results of the 1970, 1990 and 2000 censuses shows 33 developing countries with a high index of concentration of landholding (Gini coefficient at 0.5 and over), 15 of which were South American, where the concentration index was 0.7 and over. The available results of the census of 2000 show that this index in Latin America has *not* improved and, with the exception of Panama, Dominica and Honduras, it has even worsened between the 1970s, 1990s and 2000 in Brazil, Paraguay, Uruguay and Venezuela, all having a very high concentration at a Gini coefficient of 0.8 and 0.9. Their degree of land concentration is the highest in the developing countries that have conducted agricultural census. As shown in Table 3.4, concentration has also increased between 1980 and 2002 in Egypt, Bangladesh, Pakistan, Philippines and Turkey, but not in Thailand. Interestingly, the stable index of inequality in land distribution in Thailand at the moderate level of 0.46 is due primarily not to redistributive land reform, but to rapid expansion in irrigated area through the government's substantial investment in irrigation

Table 3.4 Changes in the concentration of size distribution of landholding in 23 developing countries, 1950–2004

Countries ranked in descending order of number of agricultural population per hectare of arable land in 2002	1980	2002	Land concentration index – Gini coefficient				
			1950s	1960s	1970s	1980–84	2000–04
Sri Lanka	3.4	15.8		0.66 (1960)	0.51 (1973)	0.62 (1982)	0.38 (2000)
Egypt*	7.8	13.9	0.61 (1950)	0.38 (1965)	0.45 (1975)	0.43 (1984)	0.67 (2000)
Bangladesh	7.2	13.2		0.47 (1960)	0.43 (1977)	0.55 (1984)	0.62 (1996)
Costa Rica	1.2	7.3	0.79 (1950)	0.78 (1963)	0.83 (1973)		
Indonesia	4.1	6.0		0.62 (1963)	0.72 (1973)		0.46 (1993)
Nepal	5.9	6.7		0.57 (1961)	0.69 (1971)	0.61 (1982)	0.49 (2002)
Philippines	3.2	5.5	0.51 (1948)	0.50 (1960)	0.51 (1971)	0.53 (1981)	0.57 (1991)
India	2.7	4.7	0.68 (1954)	0.59 (1961)	0.64 (1971)	0.62 (1978)	0.60 (1996)
Pakistan	2.5	4.6		0.63 (1963)	0.52 (1973)	0.54 (1980)	0.60 (2000)
Korea, South	5.8	4.8	0.72 (1945)	0.39 (1960)	0.37 (1970)	0.30 (1980)	0.38 (1990)
Colombia	1.6	4.5	0.85 (1954)	0.86 (1960)	0.86 (1971)		0.80 (2001)
Kenya	6.0	4.5		0.82 (1960)	0.74 (1971)	0.77 (1981)	
Dominica	1.8	4.0	0.79 (1950)	0.80 (1960)	0.79 (1971)		0.67 (1995)
Thailand	1.7	3.1		0.46 (1963)	0.41 (1970)	0.46 (1978)	0.46 (1993)

Honduras	1.3	2.9			0.78 (1974)		0.66 (1993)
Panama	1.2	2.4	0.73 (1952)	0.74 (1960)	0.78 (1971)	0.84 (1981)	0.52 (2001)
Iraq	0.7	1.4	0.72 (1950) 0.88 (1957)		0.65 (1971)	0.39 (1982)	
Venezuela	0.6	1.3		0.94 (1961)	0.92 (1971)		0.93 (1997)
Brazil	0.5	1.1	0.83 (1950)	0.84 (1960)	0.84 (1970)	0.86 (1980)	0.91 (1996)
Turkey	0.9	0.9		0.63 (1963)		0.58 (1980)	0.80 (2001)
Paraguay	0.9	0.8		0.93 (1960)		0.94 (1981)	0.96 (1991)
Saudi Arabia	4.1	0.6			0.79 (1972)	0.83 (1983)	
Uruguay	0.3	0.2		0.82 (1961)	0.83 (1970)	0.84 (1980)	0.8/8 (2000)

Sources: Unless otherwise indicated below, the Gini coefficient for 1960, 1970,1980, 1990 and 2000 is calculated by the FAO Statistics Division, Rome, based on the results of World Agricultural Censuses. The 1950s coefficient is calculated by the author from available results of agricultural censuses except South Korea (1945), for which the calculation was based on data in *Land Reform in South Korea*, USAID, Spring Review, June 1970: Table 3; Kenya (1960) from Berry and Cline (1979: Table 3-3), index for 1970 is calculated by the author from *Statistical Abstract 1971* of Kenya, Table 84; Turkey (1963) from Berry and Cline (1979); Bangladesh (1960 and 1970) and Thailand (1970) from ILO ACRD IX 1979 'Poverty and employment in rural areas', Table VII; Bangladesh (1984) calculated by the author from the Bangladesh *Census of Agriculture* 1983–84 Vol. I, Table 5; Iraq (1982) refers to the Gini index of land ownership calculated by the author from data in A.S. Alwan 'Agrarian systems and development in Iraq', in *Land Reform*, FAO, 1986: 25, Table 2; Indonesia (1973) calculated from the results of the agricultural census. Data on number of agricultural population per hectare of arable land in 1980 are taken from *FAO 1987 Country Tables* corresponding to each country and FAO *Production Yearbook 2003* Vol. 57 for the year 2002.

Notes
Years in parentheses refer to date of census or survey. In the first column, the year 1980 is added to compare the change in density over the period 1980–2002. Agricultural population is defined by FAO as all persons depending on agriculture for their livelihood, comprising all the economically active population and their non-working dependents.
*Egypt's index refers to land ownership from several editions of the *Statistical Yearbook* except for the year 2000; it refers to landholdings.

expansion by 170 per cent between 1960 and 2002, and its distribution among poor peasants and landless workers.

The table also shows in its first column that agricultural population pressure on arable land (persons per hectare) increased sharply in 74 per cent of the total number of countries between 1980 and 2002, a trend that is alarming in low-income countries whose land concentration worsened (Egypt, Bangladesh and Pakistan). I shall return to these relationships and their implications for landless-ness and rural poverty in the case studies (Chapters 5, 6 and 7).

The increasing land concentration noted above has been associated with a rise in the minute holdings which become smaller and smaller while attempting to accommodate more and more peasants. Concomitantly, as indicated by the results of agricultural censuses, large farms in the size group of 200 hectares and over have increased in average size, particularly in South America, accounting for nearly 65 per cent of the total area of landholdings, while their share in the total number of holdings was a tiny fraction of about 5 per cent. This size group includes the multinational corporations and large plantations. This process of polarization in agrarian structures is associated with extensive tenancy, absentee ownership of land, indebtedness and rising landlessness. In the recent past, these features of polarization emerged in African countries south of the Sahara, where traditional communal land property gradually gave way to neoliberalism for privatization/individualization linked with multinationals, e.g. Nigeria, Kenya, Malawi and Madagascar.[6]

In fact, the available results of both population and agricultural censuses indicate that landlessness has increased faster than growth rates of the agricultural labour force in many countries, irrespective of the variation in their intensity of population pressure on land. This seems to suggest the tendency of mini-holders to lose their land property rights through land and credit markets, and become landless workers relying for their livelihood on uncertain wage labour and, hence, increasing their vulnerability to several uncertainties noted earlier that will be characterized later in country-specific inquiries. This continued rise in landless agricultural workers is evidence of their lack of access to lease-in or to borrowing from the credit market in order to purchase land. Significant numbers of these entrants into landlessness are likely to add to the number of the rural poor in their countries, unless their real wages increase, employment in non-farm employment expands and the number of their working days rises.

Rural women's disadvantaged access to land

In addition to institutional monopoly exhibited by way of increased land concentration in a few hands, there is another institutional failing of rural women: disadvantageous access to land. The results of recent agricultural and population censuses (and available case studies) show that the share of women in the total number of landholdings is disproportionately very small compared with their share in both the total labour force and food production and processing. A combination of bureaucracy bias and local customs has deprived rural women of having equal opportunity in the access to land, capital and technical services (extension and marketing services). This discrimination has taken place even in situations where they are household heads, owing to being widows or assuming farming responsibilities in the absence of their migrating husbands, and despite having the duty of livestock care and processing-cum-sale of their dairy products. This gender segregation in developing countries' agricultural activities, in addition to high illiteracy among rural women (between 60 and 70 per cent except in Latin America), results in efficiency loss and reduction in total output, considering that women represent 50–60 per cent of the total labour force in low-income countries' agriculture (for empirical evidence, see FAO 1983, 1999; UNESCO 1998; UNECOSOC 1999).

How inequality of land distribution affects agricultural growth?

Having comprehended the persistent, increasing magnitude of poverty and undernourishment in Chapter 1 and the chronic high inequality of land distribution earlier in this chapter, I examine in this section how this inequality affects agricultural output growth and, in turn, the value added (income) per capita workforce engaged in agriculture, an important determinant of poverty. This is a reverse of the lasting development issue initiated by Kuznets' theory (1955) on how growth affects income inequality over different stages of development (inverted U with an upward phase in which income inequality increases with growth of per capita income).

Since the early 1990s, research work has established that inequality of both landownership and income distribution is negatively correlated with subsequent economic growth, i.e. the more unequal is the distribution (higher Gini index), the lower is the rate of economic growth in subsequent periods of time.[7] The researchers have also pointed out the importance of access to other assets: credit and education. For example, Alesina and Rodrik (1994: 481) found from the analysis of 54 developing countries' data that an increase in the land distribution inequality index (the Gini coefficient) 'of 0.16 would lead to a reduction in output growth of 0.8 percentage points per year', and added that countries that introduced land reform 'had higher growth than countries with no land reform' (ibid.: 483).

I have demonstrated earlier in Table 3.2 how total and per head food production in 1960–2003 were higher in a small sample of developing countries with low inequality of land distribution than in the rest of the developing countries. This point is developed further in Table 3.5 to how inequality of agricultural landholding distribution affects the rates of agricultural output growth over the long period of 1950–2000 in ten developing countries. To simplify the comparison without a statistical analysis, the countries are grouped into countries with high inequality (Gini index over 0.6) and countries with a low inequality index (below 0.4) in the 1960s to provide sufficient subsequent time for judging changes in agricultural output growth after the 1960s.

It is apparent from the limited data given in Table 3.5 that the countries with a high land inequality index are in the Latin American and Caribbean region and

Table 3.5 Inequality of land distribution and agricultural GDP growth rates, 1960–2000

Countries according to the degree of land inequality and years of measurement		Average annual agricultural GDP growth (%)			
		1960–70	*1971–80*	*1981–90*	*1999–2000*
A. Countries with high inequality					
Argentina	0.86 (1960, 1970)	2.3	2.6	1.1	2.8
Brazil	0.84 (1960, 1970)	1.9	4.0	2.8	3.0
Colombia	0.86 (1960, 1971)	3.5	4.9	1.8	0.1
Jamaica	0.81 (1960, 1970)	1.5	0.7	0.8	1.1
Panama	0.74 (1960)	5.7	1.9	1.9	n.a.
Paraguay	0.93 (1960)	2.1	n.a.	3.6	1.9
Venezuela	0.92 (1971)	5.7	3.8	3.1	1.3
B. Countries with low inequality					
Cuba	0.21 (1978)	n.a.	n.a.	n.a.	n.a.
China	0.21 (1971)	6.2	–	6.1	3.8
South Korea	0.30 (1980)	4.5	3.2	–	4.2

Sources: Gini index is taken from Tables 3.4 and 5.5 except Cuba (from Chapter 5) and China (from IFAD 2001: Table 3.1).
Agriculture GDP growth rates are taken from the *World Development Reports*, several years, and FAO *The State of Food and Agriculture*, several years, Annex tables.

Notes
GDP, gross domestic product is the value of final products, and it comprises plants, animals, forestry and fishing.
Gini coefficient, see definition in Table 3.4, notes.
n.a., not available.

those with a low index (Cuba, China and South Korea) are those that introduced complete redistributive land reform in the 1950s, as defined and documented in the Chapter 5 case studies. In the next chapter, the variation in the productivity of land and labour by different farm size will be analysed and the sources of variations identified. For the moment, one can make general observations from the data in Table 3.5. Other things being equal, high inequality of land distribution sloweddown agricultural growth in most cases over the long period 1971–2000, while the opposite was true in countries with a low degree of inequality.

Slow agricultural growth impedes poverty reduction

But how does agricultural growth variation affect rural poverty? It depends on the changes in land distribution. If average agricultural income grows while the initial land/income distribution remains unchanged, rich landowners gain greatly and become richer, whereas the average income/consumption of all the poor may not increase enough to enable them to cross the poverty line. Hence, as initial land inequality impedes agricultural growth, so it impedes poverty reduction. From the results of his analysis of data from 47 developing countries in the 1980s and 1990s, Ravallion (2001: 1803–15) concludes, 'A high initial level of inequality can stifle prospects for pro-poor growth'. On the other hand, 'in an economy where inequality is persistently low, one can expect that the poor will tend to obtain a higher share of the gains from growth than in an economy in which inequality is high' (ibid.: 1809). Based on the findings from the review of countries' experiences in 16 case studies in Chapters 5, 6 and 7 in this book, these serious development issues will be delineated.

Summary

In this chapter, we have moved from abstraction and hypothetical grounds of discussion in Chapter 2 to empiricism of the dimension of agriculture in the dynamics of rural poverty. Agriculture continues to be the major economic activity in most developing countries; it accounts for 18–24 per cent of total GDP and 62 per cent of total employment, on average, in Africa and Asia (FAO 2004a). The discussion has been concerned with investigating how secure access to agricultural land influences households' command over their food needs, as well as individuals' nutritional status, health and poverty levels. And in famine situations, small peasants and wage-dependent landless workers are the most likely candidates for malnourishment, hunger and severe ill-health leading to death. In rainfed agriculture, prolonged drought cannot be held to be entirely responsible as several constraints are at work, including grain-traders' monopoly of power and distorted investment patterns that disadvantage irrigation expansion and an increased supply of arable land. Empirical evidence shows that, in normal times, disadvantaged groups not receiving a wage equivalent in grain are most vulnerable to malnourishment, absolute poverty and involuntary unemployment owing to seasonality-based food production and displacement by the adoption of capital-intensive technology.

This chapter has also examined briefly five agricultural structural features that are closely interlinked with poverty in rural areas in most developing countries. Of particular importance are: declining food production; the rising concentration of agricultural landholdings in a few hands; the sharp rise in agricultural population density; and the general downward trend in the ratio of actually used arable land to agricultural workforce, notably between 1980 and 2002. This is a worrying trend that has been associated with shrinking and fragmenting small landholding size and increased landlessness. The relationship between the degree of inequality of land distribution and agricultural output growth has been examined in order to understand how increased inequality affects growth negatively. These are central questions that require more investigation in the rest of the book.

Suggested reading

Ahluwalia, M. (1985) 'Rural Poverty, Agricultural Production and Prices' in Mellow, J. and Desai, G. (eds) *Agricultural Change and Rural Poverty*, Baltimore, MD: Johns Hopkins University Press.

Booth, A. and Sundrum, R.M. (1985) *Labour Absorption in Agriculture*, Oxford: Oxford University Press.

Currie, J.M. (1981) *The Economic Theory of Agricultural Land Tenure*, Cambridge: Cambridge University Press.

Dasgupta, P. (1987) 'Inequality as a Determinant of Malnutrition and Unemployment: Policy', *The Economic Journal*, 97, 177–88.

El-Ghonemy, M.R. (1968) 'Economic and Institutional Organization of Egyptian Agriculture' in Vatikiotis P.J. (ed.) *Egypt Since the Revolution*, London: Allen and Unwin.

Georgescu-Rogen, N. (1960) 'Economic Theory and Agrarian Economies', *Oxford Economic Papers* xii, 1–40.

Melville, B. (1988) Are Land Availability and Cropping Pattern Critical Factors in Determining Nutritional Standards' in *Food and Nutrition Bulletin*, Tokyo: The United Nations University.

Sen, A.K. (1981) *Poverty and Famine: An Essay on Entitlement Deprivation*, Oxford: Clarendon Press.

Other titles may be found in Notes and Bibliography.

4 Farm size and productivity

Are small farms the engine for growth and poverty reduction?

We move now from the sectoral/macro level of agriculture to the micro farm level, and address the question: where productive land is scarce and its holding is concentrated, would the break-up of large estates raise land and labour productivity, and increase employment in agriculture? Is the productiveness of small farms higher than that of larger ones? Are the textbook standardized criteria used so far in making resource use judgements adequate under different socio-economic systems? Or, to put it differently, can we judge resource use efficiency and social gains in capitalist and socialist agriculture solely in western economic terms? As I understand the current moody debate on pro-market liberalization and anti-government intervention in agrarian systems, these questions are central to the concerns of policy-makers, donors, development analysts and peasant organizations.

Foundation of judgement

I raise these questions because I have been inspired by two prudent judgements on the subject. The first comes in the succinct words of the founder of economics, Adam Smith, who remarked in his *Wealth of Nations* (1776: Book III, Ch. 2, 364):

> Compare the present condition of these (great or large) estates with the possession of the small proprietors in their neighbourhood, and you will require no other arguments to convince you how unfavourable such extensive property is to improve.

The second judgement came nearly 200 years later and reminds us that the problems in the measurability and comparability of resource use efficiency by size of farm lie chiefly in the indiscriminate use of standard terms, regardless of different social and political contexts. Gunar Myrdal, the 1973 Nobel Laureate in Economics, says:

> The very concepts designed to fit the special conditions of the Western World
> – and thus containing the implicit assumptions about social reality by which

this fitting was accomplished – are used in the study of under-developed countries. Where they do *not* fit, the consequences are serious.

Myrdal (1968: 16–17)

In respect of the criteria used to assess efficiency in the allocation of the means of production, there is not much merit in an abstract discussion of the theoretical frame as related to large and small farms. This has been well presented elsewhere.[1] Instead, I shall use empirical evidence to identify questions of efficiency, accessible opportunities to credit and technology, economies of scale and employment as they have been addressed in practice and how the conclusions were founded within the experience of specific countries. The issues to be examined and indices to be used concern the following criteria:

1   the bases for calculating factor prices of the means of production, especially in respect of the choice of shadow prices (opportunity costs) of family labour;
2   output per unit of land and per unit of labour;
3   intensity of the use of means of production and their factor combination (land, labour, capital and intermediate inputs such as seeds, manure, fertilizer, etc.);
4   employment per unit of land in terms of the number of days per year and the number of workers utilized per unit of output produced;
5   the degree of responsiveness to technical change or innovations in agriculture;
6   the unified management with family labour in the person of the resident farm operator compared with absentee owners and insecure tenancy;
7   the basis for calculating the costs of production in socialist agriculture and the reward to workers in large state-owned farms;
8   questions of motivations, incentives, social consciousness and freedom of choice.

Capitalist efficiency criteria: the western bias

Let us first consider the experience of the Anglo-Saxon countries from which most of the textbooks' criteria and indices have been abstracted. Internationally, these technologically advanced countries have achieved high output per working person in agriculture. For example, in the USA, Canada, New Zealand and Australia, typical farms managed by one resident family range in size between 500 and 2,000 acres. The high level of per capita productivity is due partly to efficient management and skilled manpower of the owner-operator, and partly to the use of a high rate of capital and accumulated technical knowledge on land that is abundant relative to labour power. In these countries, the greater the amount of capital invested in farms, the greater the need for management ability and public investment in agricultural research. (This is combined with the high quality of technical support or, as widely termed, extension service to the farmers, who typically have a good standard of education.) All these advantages contribute to higher productivity per person in agriculture. Intensive and competitive use of capital, in the form of large-scale mechanization, is thus economically justified in these countries' factor combination in agriculture.

Backman and Christensen (1967) reported that farm sizes doubled in the USA between 1940 and 1965 while the number of adult workers per farm averaged about three. They remarked that this average is the same for Indian family farms, whereas the size of individual farms in the USA was about 100 times larger than in India in 1940–65.[2] Considering that the average farm size in India was 6 acres (and very large farms 50 acres), and that the labour force in agriculture relative to the total was only 4 per cent in the USA (about 5 per cent in advanced Anglo-Saxon countries), whereas it was 70 per cent in India (58 per cent in 2005), the problems in the universal applicability of criteria for efficiency and scale of production become evident. The efficiency of large farms in advanced countries' agriculture refers to a higher return on capital and manpower (management and family labour) and is measured in terms of output per man-hour. The unity of management and family labour is an essential feature because of the inherent efficiency of absenteeism in private landownership.[3] This point is worth remembering when we examine the prevalence of absentee owners of large estates in developing countries, particularly in Latin America.

Expanding the example of the USA and India, the results of the 2000–01 World Census of Agriculture show some fundamental variations in natural endowment and factor combinations in a selection of five developed and eight developing countries. In the former, productive land is not scarce and does not constitute a limiting factor in production. The ratio of agricultural population per hectare of arable land is very low, ranging between 0.02 and 0.7 persons per hectare, while the median size of holding is large, ranging from 147 hectares in the UK to 1,307 hectares in New Zealand. In addition to the advanced skills and capabilities of farmers, this combination of low agricultural population per hectare and high median size of holdings helps to explain the relatively high number of units of capital used. The opposite is generally true in developing countries, where the agricultural labour force as a proportion of the total labour force is seven to ten times larger than in advanced or developed industrialized countries. Although these data on the labour force do not show the quality of land and its cropping pattern, the relative price of factors of production is indicative of their relative scarcity. For instance, my study (El-Ghonemy 1990b: 122) shows that the average price of 1 hectare in the USA was US$800 in 1970 and the daily agricultural wage rate was approximately US$12. Contrast these rates with Egypt's price of land and labour at the equivalent US$1,500 per hectare and US$0.5 per daily worker in the same year. There is an exceedingly wide variation in the ratios of factor prices: land is 66.6 times that of wage rates in the USA, while it is 3,000 times that of daily wage rates in Egypt. The variation has even widened in Egypt since the 1970s reaching 4,100 times the daily wage in 1975–80 and 4,375 times in 1997–98, as shown in Chapter 7. This variation cannot be explained solely in economic terms. We cannot ignore the historical context, the institutional framework of property rights, the degree of imperfection in factor markets and custom-determined production relations, within which resources are allocated and their economic returns determined.

Because cultivable land is scarce and expensive in many developing countries, the economic response of their farmers is to maximize the return on output per unit of land. Thus, the *intensity* of land use is very high, particularly in irrigated areas [for example two or three crops a year are produced from one feddan (acre) in Egypt]. In developing countries (LDCs), yield-increasing and labour-using fertilizer is adequately applied in general terms, while labour-saving capital equipment, such as tractors, cannot be justified in economic or welfare terms. In LDCs, tractors are mostly concentrated on large farms and highly commercialized plantations. The use of traditional farm tools by small farmers (hoes, hand sickles, buffalo- or ox-driven ploughs and threshers) instead of heavy tractors and combine harvesters is not a manifestation of irrational economics or primitive methods. On the contrary, faced with institutional constraints in the highly imperfect credit market and low demand for labour by the capital-intensive non-agricultural sectors, these thriving, hard-working peasants are indeed rational.

This resource use variation according to the size of landholdings (farms) is manifested in the results of the recent World Census of Agriculture (FAO 2004a: Table A8). The results show that developing countries with a high density of agricultural labour force on cultivable land and a moderate degree of inequality in terms of the Gini index, presented earlier in Table 3.5, have a very low ratio of tractors per 1,000 hectares: Bangladesh, 0.7; Indonesia, 4; the Philippines, 2. On the other hand, those with a very low density and a high degree of land concentration use a much higher ratio of tractors per 1,000 hectares: Brazil 13.7 and Venezuela 20. A similarly high ratio of tractors per land unit is observed in advanced countries having low density and very large average size of landholdings: Canada 16 and the USA 27.

At this point in the discussion, it is important to indicate that, if social environment, intrafamily relations and institutional barriers are ignored, the habitual use of criteria for measuring efficiency under unrealistic assumptions of perfect competition, zero transaction costs and complete information availability is of little economic meaning. We are also deluded by the consequential prediction of outputs under conditions of certainty (in the neoclassical sense of a farm operating with complete technical knowledge and unlimited capital). It is this set of western-biased criteria of economic relations that is still used by some economists and business agents in agriculture to advocate large farms and commercial plantations, including the multinational corporations (MNCs) in developing countries. The advantages of large estates, according to this view, are as follows: they use technical knowledge, competent management and skilled manpower as in modern industry; the level of education among the landlords provides them with entrepreneurial skills; and large estates increase the marketable surplus, domestic saving and capital formation. In addition, and equity considerations aside, large estates are considered to be pioneers in applying technology, which they diffuse among small farmers; they release agricultural labour power for urban and industrial development; and they provide efficient marketing services and a high quality of agricultural products, and so on.

Based on this reasoning, those in favour of large estates call for an agricultural development strategy that concentrates scarce resources in highly commercialized and profit-motivated large farms, regardless of the distributional consequences of inequality and poverty. Different adjectives were given to this pattern of agricultural growth such as 'modern', 'dynamic', 'progressive', 'capitalist', 'neoliberal' and 'bimodal' (in the case of a dualized agrarian structure). We shall examine in Chapter 5 the efficiency and equity impacts of dualized agrarian structure (land reform and non-reform subsectors in the same country) in partial land reform case studies.

Socialist efficiency criteria: state farms

The Anglo-Saxon doctrine that 'to be efficient one must be very large and intensive in capital investment' has also been a major feature of state-owned and -managed farms in several countries, since their beginnings in the former Soviet Union in the 1920s. These large farms have to be seen as an integral part of the economic structure, centralized authority of the state, nationalized land property and collectivized farming. They also have to be studied within moral and material objectives, and the incentive systems of specific countries over defined periods of time.

The meanings of 'employment', 'profit' and 'incentive' vary greatly, not only among socialist countries (referred to in ordinary parlance as communist countries), but also within countries during periods of time. For instance, concepts in China under Mao Tse-Tung (1949–76) differed from those of Zhao Ziyan Deng, initiating readjustment and management of the agricultural economy in 1979 as part of national economic reform. State farms in Cuba are the most extensive in relative terms in all countries, 80 per cent of the total agricultural land area in the 1980s. They have remained although the continuing Cuban leadership of Fidel Castro Ruiz has introduced changes that he called a process of prominent rectification of socialist errors and negative tendencies, including workers' participation in the management of state farms and launching the food security programme to reduce dependency on imports (Deere *et al.* 1998). Therefore, it is difficult, if not impossible, to generalize on the performance and efficiency of state farms. Judging from my field studies in Russia, Hungary and Mozambique, they are diverse in organization and role in the agrarian economy, not only between countries but also between one state farm and another in the same country. There are still, however, some common features which could be broadly outlined. They are centrally planned and are giants in size and in their share of capital stock, and many of them form large agro-industrial complexes. In addition, many state farms specialize in the production of one commodity and have widespread use of technology. They provide the state with command over the production and marketing of essential food and cash crops for export, thereby ensuring the supply of food to the urban population (including armed forces and civil servants). Their generation of employment is considerably expanded when the production of raw material is integrated with processing in a complex of agro-industry. Lastly, although centrally managed by skilled staff, they suffer from heavy bureaucracy.

In order to appreciate the large scale of operations of state farms until the late 1980s, a few examples may help. According to my study conducted in 1984, in the former Soviet Union (Russia), state farms accounted for 106 million hectares, representing 66.4 per cent of total agricultural land in 1978. They employed 11.3 million employees. According to the findings of a wider study, state farms in China account for a much smaller proportion of land, only 4.3 per cent of total agricultural land in 1982, but they employed 4.8 million – almost half the number of state farm employees in the former Soviet Union. Nevertheless, the area of the state farms in China was substantial – 4.4 million hectares. In the 1980s, the proportion of state farms' acreage to total agricultural land varied in other countries: 85 per cent in Cuba, 5 per cent in Vietnam and Ethiopia and 4 per cent in Mozambique. On average, the size of each state farm is considerable, ranging from 2,000 to 16,000 hectares.[4]

The role of these important institutions in the national economy has to be seen against the historical background of a long colonial history (e.g. Cuba, Mozambique and Vietnam) or semi-feudal agrarian economies (e.g. China, Ethiopia and Russia). Large farms previously owned by absentee landlords or foreign-owned plantations were converted to state-owned and -managed farms. With the exception of Stalinist Russia, agriculture has been given absolute priority in development strategy. The priority in China was expressed by Mao in the 1950s, 'Agriculture is the foundation of National development from 1960 onwards' (White 1983: 12). However, since the introduction of several types of economic reforms in the 1980s and 1990s, there have been such dramatic shifts, and rethinking of economic management within agriculture in many countries, that it is difficult to catch up with these changes. However, empirical evidence on current changes in production incentives and agricultural land property/use rights will be examined in the case studies presented in the next three chapters.

Despite being fortunate to visit and study several countries with socialist agriculture between 1960 and 1991, I find measuring efficiency of resource use in state farms in economic terms difficult for a foreign analyst, such as myself, particularly with respect to the difference in cost accounting procedures.[5] The difficulty also rests on whether the state farm is examined as an autonomous enterprise or just an integral unit of a wider economic and political structure. Procedures also vary according to the relation of central planning to the 'market mechanism'. Wages, salaries, bonuses and prices of the means of production and output are administratively and politically determined in many socialist countries. It is not an uncommon practice for the large number of employees and farmworkers to have guaranteed salaries and wages, regardless of the actual level of productivity. In cases where state farms fulfil the established target of gross output, all staff and workers are usually paid bonuses. While employment of the large number of management staff and workers is guaranteed by the state, the lack of clearly defined material incentives to work harder presents problems.

In Cuba, for example, reliance on conscientiousness combined with ambiguous material incentives had disastrous consequences for the newly established sugar cane state farms in the 1960s. However, when the macro-indicators of performance in the agricultural sector as a whole are used, the performance of the socialist countries appears to be as good as those with private property market economies. In fact, some socialist countries (e.g. China, Cuba and North Korea) performed even better. Two such overall indicators were used by Keith Griffin (1986: 173): the annual growth rates of both agricultural production and labour productivity during 1970–80. In his words, 'Hence there is nothing in these data to suggest that the performance of countries with communal tenure systems is inferior to those with individual tenure systems'.[6] Perhaps these macro-indicators of performance can serve as an approximation of the performance of state farms in countries where the resulting value added is the major source of total agricultural GDP, such as in the case of Cuba, where the share is 80 per cent according to Ghai *et al.* (1988). Such studies are inhibited by a multitude of factors, some of which have already been mentioned, such as material and non-material incentives, centralization and decentralization of management, and strict hierarchical chains of command. To these we can add politicization of state farms, bureaucratization of agriculture and labour utilization, linking or de-linking the planned economic activities with or from the market, and the different definitions used by state farms with regard to what are 'productive' and 'non-productive' services.

But terms such as centralization or decentralization of management are ambiguous and require clarification from country-specific empirical studies. For instance, a study on Cuban rural economy by an international team of experts brings some of these factors to light. It judges the performance of state farms favourably based on official figures of crop yields, which show an increase of about 4 per cent per annum between 1975 and 1983. According to the findings of this study, the objective of the management system, since about the mid-1970s, has been to promote economic efficiency. Key features include economic accounting at the enterprise level, prices based on cost of production, decentralized decision-making and material incentives to boost productivity, efficiency and equality of output. However, the Cuban economy's being centrally planned has naturally determined the limits of decentralization and the devolution of decision-making. Input and output prices are fixed by the state, and the wage rates and the payment systems are also laid down by the state (Ghai *et al.* 1988: Ch. 7).

This good record of state farms in Cuba is not matched in Mozambique, according to my own study during April and May 1984 (FAO 1985b). Three state farm areas were studied: Capo Delgado in the north, Zonas Verdes in the south, and Chekoe in the west. Although representing 4 per cent of agricultural land, state farms cover 150,000 hectares of the best, mostly irrigated, land. They produce strategic crops for export, food consumption in rural areas and for agro-industries (cashew nuts, maize, rice, cotton, tea, sunflowers and cocoa). Following independence in 1975 after five centuries of Portuguese rule, the country's leadership assigned state farms high priority in the investment of a large share of two scarce resources: foreign exchange and technically qualified staff. Nevertheless, expectations were not met. The three studied farms experienced serious management problems of inadequate technical personnel, a lack of professionally skilled managers and a high degree of politicization of resource use management. I also found that product prices or daily wages were centrally determined by bureaucrats in the capital city, Maputo, and supplies of agricultural equipment and spare parts for crop processing factories were seriously inadequate. These constraints are manifested in the steady decline in the volume of agricultural exports by 8 per cent per annum during the period of the study, 1976–83. Moreover, cotton and cocoa oil exports fell by 26 per cent during the same period. As shown in Table 4.1, the marketed sunflower and cocoa fell sharply between 1981 and 1983, and cotton output dropped by 12 per cent during the period 1977–83. Compared with the communally cultivated land (cooperatives known as the peasant sector), there is a slight variation suggested by the data on Zonas Verdes in Table 4.1. This was also noticed in Chekoe state farms growing maize and rice. Productivity per hectare in their state farms averaged 2.5 tons of maize and 3 tons of rice compared with 1.5 and 2.5 tons respectively in the co-operative/peasant sector.

Empirical examination of efficiency in developing countries

Questions of the advantages, merits and economic performance of large estates in developing countries with private property market economies are fundamental. They should, therefore, be left to empirical evidence. We shall use for this pur-

Table 4.1 Mozambique state farms' marketed crops, 1981–83

| | | Zonas Verdes | | | | Capo Delgado* | |
| | | State farms | | Cooperatives | | | |
Crops and years		Tons	Indices	Tons	Indices	Tons	Indices
Maize	1981	33,789	100	1,704	100	3,541	100
	1982	47,477	141	1,458	86	8,036	226
	1983	27,232	81	785	46	5,026	142
Rice	1981	25,594	100	1,407	100	377	100
	1982	38,677	151	979	70	527	139
	1983	15,022	59	546	39	1,139	302
Sunflower	1981	3,285	100	509	100	934	100
	1982	933	28	245	48	577	62
	1983	602	18	152	30	383	41
Cocoa	1981	15,000	100				
	1982	12,710	85				
	1983	9,796	86				

Source: Compiled by the present author during his visit to the State Marketing Organization, AGRICOM E.E. in Maputo.

Note
*No disaggregated data for state farms and cooperatives (collectives in Aldeas Comunales). Data before 1981 were not given because AGRICOM was established in that year.

pose two sets of evidence: a cross-sectional analysis and case studies of farms or households in certain countries listed in notes 7 and 8. The discussion consists of four sections on economic issues deducted from our analysis of the results of the empirical studies:

1 productivity of land and labour by farm size;
2 economics of scale as an indicator of the economic performance of large and small farms;
3 responsiveness of farms to technical change according to their size of holdings; and
4 the combined impact of large farms' neglect of food production and the labour market segmentation.

Productivity of land and labour

The wide range of findings uncovered by these penetrating studies confirms that, with slight variations, output per unit of land in all countries declines systematically with a rise in farm size. Physical output and value of gross output per unit of land is consistently much lower in large farms than in small farms. The range of variation is very large in land-abundant and sparsely populated countries (e.g. South America) compared with land-scarce countries with population pressure on the land (e.g. Egypt, India and Pakistan). Large estates (haciendas) in Latin America, which constitute 50 per cent of total farm land in the seven countries studied by the inter-American committee for Agricultural Development (CIDA), are characterized by underutilization of land, widespread absenteeism and a shift towards pasture for raising livestock. The study asserts that this manifestation of the inefficient use of the resources of large estates was inconsistent with their high management capacity and their advantageous position in access to credit, technical assistance and in water supply for irrigation (Barraclough 1973: Ch. 2).

There is a consensus among the researchers in their cross-sectional analysis[7] that labour utilization per unit of land (measured in terms of man-days per year and number of workers per unit of output) is considerably lower in large estates than in smallholdings. This was especially apparent in countries with a high degree of unequal distribution of land. Moreover, the studies clearly show that labour intensity per unit of land is positively correlated with land use intensity and negatively correlated with farm size (land use intensity means the ratio of cropped to total farm area). This relationship provides part of the explanation for the inverse relationship between output and the size of landholding. Cornia's (1985) cross-sectional analysis of 15 countries' data found that there is an excessive amount of labour crowded in small farms, implying a wasteful use of labour power and a declining marginal productivity of labour input. This situation coexists with underutilized land in large estates in the same country, suggesting that the peasants are trapped in their small farm sector. The three studies listed in note 7 have shown that the imperfection of capital markets has aggravated rural underemployment. As large estates have preferential access to cheap credit (low interest rates and highly subsidized machinery), they substitute machinery for labour. They deliberately choose lower labour to land ratios than would result from an imperfect labour market alone.

Undoubtedly, the available series of cross-sectional analyses provides us with interesting broad indicators of resource use and output related to the size of land-holdings. Still, the authors face data limitations inherent in all comparative analyses of the situation in a large number of countries. Important among these are:

1   the type of soil and productive capacity of land area (which differs from farm to farm and from one country to another);
2   the costing of family labour inputs;
3   the choice of shadow prices (opportunity cost) of the means of production in judging total social productivity and potential gains;
4   capturing the effects of absenteeism among landowners on the application of technology and on day-to-day operational decisions; and
5   the differences in the institutional contexts of production relations.

It is because of these limitations and the diversity of situations that case studies of specific countries are more meaningful, especially when they are based on large samples of agricultural households. They usually accommodate the above-mentioned variations across a range of farm sizes. The sets of data they provide, when combined with the production function for a specific situation, make their findings on factor proportions and production relations more relevant than a cross-sectional analysis.

We are fortunate to have a number of such studies in several developing countries.[8] They examined in depth the fundamental issues of the relative technical and economic efficiency of large- and smallholdings under different conditions of natural endowments and social systems. Some of them examined these issues in the same country within a 3- to 20-year interval to find out whether there was any change in the pattern of relations. Examples of such studies are: Sarjit Bhalla's analysis of data on 1,772 agricultural households collected three times over the period 1968–71 by India's National Council of Applied Economic Research; William Cline's analysis of data collected in 1973 on Brazil, which he compared with his earlier 1963 survey of seven major districts; Hayami and Kikuchi's surveys of two villages in the Philippines carried out in 1966 and 1976/7; Michael Henry's case study of the economies of scale in rice production in Guyana in 1974 and 1984; El-Ghonemy's sample survey of 611 land reform beneficiary households in 1973 after 20 years of his macro-agrarian study conducted in 1953 (in El-Ghonemy 1990a: 234–5); and Paul Collier and Deepak Lal's analysis of Kenya's Integrated Rural Surveys 1974–79 and the 1978 Labour Force Survey.

There are other country studies including Keith Griffin's examination of the situation in Ecuador and Morocco, Doreen Warriner's 1969 field findings in Brazil, Chile, Venezuela and pre-land reform Iraq, as well as my own study of resource use and income in pre-land reform Egyptian agriculture (1935–51) and in three land reform districts (1973). In addition, studies on the labour market and technology by size of landholdings in Egyptian agriculture were conducted by Commander and Hadhoud in 1984. Thiesenhuesen (1995) also studied land productivity by size in Brazil and Mexico.

Although the chief concern of these comprehensive studies varies, all in all, they illuminate our search for the dynamics of rural poverty. They examined the pattern of production relations between large and small farms and the implications for agricultural/rural development. They also provided explanations of the economic and technical causes and the institutional barriers inhibiting the small farmers' access to credit, technical knowledge and water rights for irrigation for small farmers. In addition, Bhalla, Hayami and Kikuchi, El-Ghonemy and Henry investigated the question of whether the pattern of factor combination established in their previous studies continues after the introduction of high-yielding varieties and non-crop sources of income in the 1960s and early 1970s and the associated investment activities required by technical change. Henry went on to consider the effects of a continuing pattern on production, employment and land tenure relations, and El-Ghonemy examined the changes in non-land/non-crop sources of Egypt's land reform new owners' income between the two studies, 1953–73.

The findings of these country-specific case studies confirmed the conclusions reached by the cross-sectional studies in respect of lower output and labour input per unit of land among large farms compared with smallholdings. In his analysis of the data set on India, Bhalla ascertains that these relationships remain significant even after differences in land quality are allowed for (in Berry and Cline 1979: 154). He remarks that, although imperfection in all factor markets is responsible for these relationships, imperfection in the labour market is the most important factor (in Berry and Cline 1979: 172). The studies on the Philippines and Kenya (Central Province) showed an association between absentee landlords and land concentration. In the former, more than 70 per cent of 'South Village' land in 1977 was owned by five large landlords who lived in Manila (Hayami and Kikuchi 1985: 135). In Kenya, 'a considerable amount of land purchase has been undertaken by absentee *urban* high-income groups. This has increased land concentration and worsened the imbalance in factor proportions between large- and smallholdings (Collier and Lal 1986: 132). In Chapter 6, we shall return to this land market approach followed in Kenya. Thiesenhuesen's studies on Brazil and Mexico showed that the land productivity of small farms is at least twice that of large farms, owing primarily to double cropping and no or minimum fallowing (1990).

In my pre-land reform study on resource use in Egyptian agriculture (El-Ghonemy 1953), small farms of less than 5 acres, representing 81 per cent of total holdings, met 95 per cent of their total labour requirements from their family labour, using only 2 per cent of threshing machines, 3 per cent of total tractors used and 6 per cent of irrigation pumps (which were shared among agricultural cooperatives). At the other extreme were large landholders of 100 acres and over who represented less than 1 per cent of the total number of landholders. These landholders had access to credit and technical knowledge, and they often employed qualified managers. They often relied on hired labour and less intensive land use, as they owned 80 per cent of the tractors, 83 per cent of total threshing machines and 35 per cent of irrigation pumps (most with high horsepower). A later study conducted during 1984 shows that smallholdings of less than 1 acre had higher productivity (yield per acre) than holdings of larger areas. (It should be noted that the maximum ownership of land prescribed by land reform is 100 acres per household.) When the value of land productivity was computed, using regression analysis, the results show that productivity rises for farms up to 10 feddans (acres) but falls off for the larger farms. The resource allocation among the crop rotational combination in small farms is believed to be responsible for this variation (Commander and Hadhoud 1986: Ch. 8, Table 8.E). Calculated average productivity per labour hour (physical units in kilograms) for wheat and cotton is higher for small farm sizes below 10 acres than above that size (ibid.: Table 8.5). As virtually all farms in Egypt are irrigated by Nile water, and material inputs such as fertilizers and insecticides are accessible to farms of all sizes at a heavily subsidised price, the difference in productivity can be attributed to intensive family labour on small farms.

Economies of scale

When the prices of production factors (inputs) are included in calculating the ratio between the value of total output to the value of all factor inputs, the economic efficiency of large farms appears to be lower than that of small farms in the case studies of India, Brazil, Guyana and the Philippines. This is an important indicator of the economic performance of large estates, and it rejects the claim of their dynamic superiority in production and employment. As Professor Cline states in his study of Brazil, 'the large farm sector uses its available land inefficiently from the standpoint of the economy as a whole' (Berry and Cline 1979: 58). Likewise, the study of the Philippines concludes that 'there are no significant differences in the economic efficiency of large and small farms' – overall, 'there is no evidence that large farms were most efficient technically than the small' (Hayami and Kikuchi 1985: 144). Henry, in his sample survey of rice production by size of holding in Guyana, found, through a rigorous statistical analysis of the data, that the large farms were economically inefficient beyond 15–20 hectares under the same assumptions of economies of scale. Referring to his country, Guyana, Henry says, 'In this case the price we pay for efficiency is a preponderance of large farms along with the attendant problems of socio-economic development' (Henry 1986: 11).

Technical change

We next turn to the responsiveness of farmers to technical change according to farm size. The widely held view that large estates are superior to small farms in technical change has proved to be generally false. This was clear from the hard evidence brought by the studies cited above and many others on the adoption of high-yielding varieties (HYV) and related technology.[9] Farmers appear to adapt to new processes of production despite their disadvantageously low initial endowment of land and capital and the high cost they must pay for credit. This positive response by small and large farms to technological change has been documented in many field studies: Azam and Khan on Pakistan, Bardhan and Sen on India, Bhalla on India, El-Ghonemy on Egypt, Collier and Lal on Kenya, Taussig on the South of the Cauca Valley in Colombia, etc.

Consider one example. In his study on India, Bhalla indicates that the percentage area of HYV of rice and wheat was higher in farm sizes below 5 acres than in farms above 25 acres in 1968 (14.6 per cent compared with 11.1 per cent respectively). But large farms expanded their area faster than small farms because the latter suffered from credit constraints, particularly for irrigation capital, which is essential for this type of technical change. In terms of the proportionate number of people adopting technology, the difference was slight (Berry and Cline 1979: Tables A-16 and A-17).

Moreover, hard evidence was brought by the case studies of Guyana and the Philippines in respect of changes in the production process resulting from the adoption of new rice technology (seed, fertilizers and weeding). There was no difference between large and small farms. The only difference, as expected, was in mechanization (harvesting and threshing) introduced by large farms. Despite the high quality of management in large farms, the Philippines study states clearly, 'With *similar* input use, there was little difference in paddy yield per hectare on large and small farms' (Hayami and Kikuchi 1985: 144). Collier and Lal indicate that Kenyan small farmers, despite severe credit rationing by commercial banks and marketing cooperatives, were able to meet their household food needs and to grow modern hybrid maize in addition to pyrethrum, coffee and tea as grown in large farms (Collier and Lal 1986: Table 5.4).

Food production neglect and labour market segmentation in large farms: the case of Kenya

In several Latin American and African countries as well as the Philippines, large farms including commercial plantations and MNCs (e.g. Del Monte, Brooke Bond, Findlay and Unilever) dominate the agrarian structure within a capitalist system. To illustrate this influence, the situation in Kenya is briefly examined. The results of post-1974 censuses of agriculture and the findings of available studies indicate a striking dual agrarian structure. Thereby, the large farms in the average size group of over 1,000 hectares (mostly foreign owned) represent almost one-third of the total number of landholdings; each has 3,700 hectares. See Bates (1981: 93–5) on their policy influence and market power. In contrast, native small farms constitute nearly two-thirds of the total number with an average size of 2 hectares each.

This contrasting structure has shaped resource use in Kenyan agriculture. The large farms sector grows high-value crops for export (tobacco, tea, coffee, sisal and medicinal plants), while intensive food cropping dominates the small farms sector (millet, maize, sorghum, potatoes and cassava). In the latter sector, landlessness persists, and the prevalence of poverty and undernourishment is very high (47–52 per cent and 33 per cent respectively). With population rising at an annual average rate of 2.4 per cent and the rationing of credit supply in the small farms sector, Kenya has been unable to meet its growing demand for food, while the food productivity per person of total population is falling. Accordingly, cereal imports increased substantially from 27 per cent in 1987 to 40 per cent of the total agricultural imports in 2001.

With regard to the influences of the large farms' substantial market power on labour use and wage rates, the rich landlords and MNCs succeeded in exempt-ing their sector from the application of minimum wage laws, thus depriving the hired workers of substantial equity gains in addition to their loss of employment opportunities, owing to the adoption of labour-saving technology, while the small farms sector is suffering from capital rationing.

To conclude this Kenyan case, we say that, despite the influx of private for-eign capital and managerial skills into the large farms and MNCs sector, the high degree of inequality in land distribution combined with the defective institutional framework of Kenya's agriculture have resulted in a grim picture of rural develop-ment not only in physical agricultural output, but importantly in the advance of human capabilities, as suggested by the data in Table 4.2. Production of crops and livestock as well as food productivity has fallen between 1980 and 2003. Alarmingly, during the same period, average annual productivity in agriculture, i.e. value added per working person in agriculture in real terms, also fell from US$184 to US$148. Given the high share of agriculture in total GDP (25%), its poor performance has pulled down the national income from average annual

Table 4.2 Rural development indicators in Kenya, 1979–2003

Indicator		*Year/period*	
1	Agricultural production average annual growth	1980–90	3.3%
		1991–2003	2.5%
2	Agricultural production in real terms per working person in agriculture	1989	US$184
		2001–03	US$148
3	Rural poor as a percentage of total rural population	1994	49
		1997	53 (59)
4	Undernourished people as a percentage of total population	1990	10.6%
		2001	11.5%
5	Mean years of schooling (1999) Rural 5 Urban 8		
6	Infant mortality per 1,000 live births	1970	96
		2001	78

Sources: 1 and 2, FAO *Statistical Yearbook* (2004: Vol. 1); 3, *World Development Report* (2006: Table A1); 4, *The State of Food and Agriculture*, FAO, 2003–04, Table A2; 5 and 6, *Human Development Report*, UNDP, 2003.

Note
In item 3, rural poverty is measured by national household survey. The percentage in brackets is the World Bank's measurement of the poverty line at US$2 per day per person.

growth of 4.2 per cent in 1980–90 to a low average of 1.8 per cent in 1990–2003. What is disturbing is that undernutrition and rural poverty prevalence worsened between 1990 and 2001.

There is no merit in continuing with other examples of empirical evidence on resource use by farm size. It is clear from what we have already presented that in risk bearing, in food production, in their responsiveness to technological change and in economic efficiency or resource use generally, large farms are not superior to small farms, despite the latter's formidable institutional obstacles.

Summary

The chapter has addressed the central policy question in the current neoliberalism debate at grass roots, national and international levels with regard to potential economic and welfare gains from breaking up large estates for the benefit of the rapidly increasing number of agricultural landless workers and poor peasants. The relevance of this question to current concern is critical where land is scare and its ownership is concentrated in a few lands. Or, to put it differently, is the loudly voiced myth valid that large farms per se are superior to small ones in resource use efficiency, and that they are necessary for dynamic rural development? But what rural development is referred to in this myth? It has been widely accepted in the literature that it is anti-poverty participatory change, in which opportunities for equal land access are prominent (see Chapter 2).

In addressing this central policy question, the sufficient and reliable empirical evidence presented suggests four concluding remarks.

First, within each country's specific ideological circumstances and historical context, it is necessary to distinguish between private property large farms in the market economy (including commercial plantations and multinational farms) and state-owned and -managed farms. Within this differentiation, we can neither use economic criteria solely nor apply the standardized western efficiency measurement. Instead, it is necessary to include the variation in production and earning incentives, pricing and cost accounting as well as the producer's freedom in decision-making (centralized in the government's or the entrepreneur's own decision with regard to freedom to employ resources, to combine them for raising output and reducing costs and freedom to trade).

Second, land concentration in a few hands is a major mechanism for exercising power-generating institutional monopoly and labour market segmentation in developing countries, with capitalist agriculture. The neoclassical and neoliberal assumptions of perfect market competition, zero transaction costs and complete access to information and technical change are unrealistic.

Third, the sets of empirical evidence brought forth – from carefully conducted studies in several countries – by a number of scholars who are eminent in the subject of rural development reveal that large estates, in general, are inefficient in resource utilization, total productivity of factor combinations and in intensity of both land and labour use. This does *not* imply, of course, that *all* large farms are inefficient *everywhere* in developing countries.

Lastly, all the studies cited in this chapter either suggest land redistribution or recommend specific redistributive land reform actions as an anti-poverty device. For instance, the comprehensive study of 15 developing countries[10] states, 'the higher yields observed in small farms are ascribed to higher factor inputs and to a more intensive use of land. Therefore, conspicuous labour surpluses exist, the superiority of small farming provides solid arguments in favour of land redistribution, thus contributing in a decisive manner to the alleviation of rural poverty' (Cornia 1985: 513) 'because of the demonstrated superiority of small vis-à-vis large farming, land redistribution, if thoroughly implemented, would have immediate beneficial effects in terms of output growth, enhanced income distribution and, as a result, alleviation of rural poverty' (Cornia 1985: 532). Similarly, Cline, Griffin and Henry recommended in their studies specific actions for land redistribution in Brazil, Morocco, and Ecuador and Guyana respectively.

Suggested reading

Backman, K. and Christensen, R. (1967) *The Economics of Farm Size, Agricultural Development and Economic Growth*, Ithaca: Cornell University Press.

Bardhan, P.K. (1984) *Land, Labour and Rural Poverty – Essays in Development Economics*, Delhi: Oxford University Press.

Barraclough, S. (ed.) (1973) *Agrarian Structure in Latin America – A Resumé of CIDA Land Tenure Study*, Lexington, KY: Lexington Books.

Berry, A.R. and Cline, R.W. (1979) *Agrarian Structure and Productivity in Developing Countries*, Baltimore, MD: Johns Hopkins University Press.

Bruce, W. (1972) *The Market in a Socialist Economy*, London: Routledge & Kegan Paul.

Cornia, G.A. (1985) 'Farm Size, Land Yields and the Agricultural Production Function: an Analysis of Fifteen Developing Countries', *World Development*, 13, 4.

Dorner, P. and Kenel, D.R. (1971) 'The Economic Case for Land Reform. Employment, Income Distribution and Productivity', *Land Reform, Land Settlement and Cooperatives*, 1, 59–68.

Ghai, D., Kay, C. and Peek, P. (1988) *Labour and Development in Rural Cuba, An ILO Study*, London: Macmillan.

Hymer, S. (1971) 'The Multinational Corporation and the Law of Uneven Development' in Bbagawa, J. (ed.) *Economics and World Order*, New York: Macmillan.

5 Case studies I

State-administered complete land reform

In this and the following chapters, we advance the empirical discussion to country-by-country production and distributional consequences of land reform policy choice. The case studies are presented with two chief aims in mind. The first is to understand how the varied agrarian systems that existed before land reform explain the policy choice of different property rights arrangements. The second is to explore how the policy choice under different ideological preferences determines the secure accessibility of agricultural landless workers and poor peasants to poverty-reducing opportunities within the process of rural development. This secure accessibility is studied within the created system of incentives and the different types of agricultural production organization.

To maintain a balance of presentation, this chapter reviews policies implemented in China, South Korea, Iraq, Cuba, Mexico and Tunisia, while those of Egypt, India, Iran, the Philippines and Syria are examined in the next chapter. These countries have been selected for having different political ideologies, extent of property rights redistribution, density of agricultural population relative to cultivated land, social structure and historical contexts.[1] To capture the dynamic production and distributional changes, particularly the resulting changes in poverty levels, the review covers a post-reform period of 30–40 years.

Classification into complete and partial land reform

To satisfy the aims of this review, countries are grouped into two broad categories according to the scope of change in land distribution. This simple classification is made despite the existence of a variety of forms of agricultural production organization associated with property rights in land. The countries are classified as having:

1 a complete land reform policy; and
2 a partial policy, dividing the agrarian system into reform and non-reform sectors.

For the purpose of our study, a land reform policy is *complete* if it meets the following conditions:

1   The beneficiaries have direct access to individual or collective landownership representing at least two-thirds of total agricultural households.
2   All or at least two-thirds of the landless peasants are absorbed, leaving none or a small fraction as agricultural landless workers.
3   The area of the redistributed cultivable land amounts to over half the total.
4   There is sustained command over food in terms of per person food production and nutrition status.

With regard to *partial* land reform, redistributive requirements are relative to the four conditions listed above. This means a lower percentage of new landowners to total agricultural households and a correspondingly smaller proportion of redistributed cultivable land area. Differences in the scale of reform are traced to the level of size ceiling established on individual landownership relative to the average size of redistributed units. Consequently, partial land reform is likely to leave a substantial section of landless agricultural households who remain as either small tenants or wage-based workers. To understand our classification, a sample of 22 countries with these specifications is presented in Table 5.1 according to the criteria suggested. The most recent land reform policy in Bolivia, proclaimed in May 2006 – for the redistribution of nearly 2 million hectares among the indigenous rural population – is not included for lack of relevant information (see Appendix C).

In this empirical review, no preferences are implied for one approach over the other. Rather, the intention is to show, as objectively as the data permit, how the scope of land reform policy influences the distribution of income or consumption, productivity per agricultural worker and the prevalence of rural poverty. We also attempt to identify the dynamic forces operating in the national economy which are exogenous to the institution of property rights in land and which, following land reform implementation, tend to stabilize or disequalize the pattern of rural income distribution over time and, in turn, poverty levels (e.g. pricing policy and non-farm employment opportunities).[2] In both categories of land reform, the sustainability of the initial redistributive gains depends on the sequential provision of complements of production inputs and institutional arrangements for borrowing credit and marketing products to replace those abolished.

As paramount aims in development, improvement of nutritional standards, life expectancy and educational levels is given special emphasis in the review for improving human capabilities and contributing to increased productiveness of land reform beneficiaries and indirectly to agricultural gross domestic product (GDP) growth.

Complete land reform

China, South Korea, Iraq and Cuba are the countries selected for review under the category of complete land reform policy. They are followed by Mexico and Tunisia, which are considered quasi-complete; despite being an approximate or relative classification, their criteria are closer to complete than to partial land re-form (Table 5.1).

China

The rapid transformation of the Chinese agrarian economy over the last half-century has been of great interest to development analysts, practitioners and international organizations. Our concern is to probe the extent to which the reformers' choice of the institution of property rights in land, together with other means of production, has shaped agricultural growth and equity, thereby affecting the prevalence of rural poverty. The fundamental changes initiated during the series of reforms (1948–78) are best viewed against an understanding of the characteristics of the initial conditions of agrarian structure.

Pre-1948 agrarian conditions

Our assessment of pre-1948 land reform conditions is based on: Buck's survey of farms in North and East-Central China during 1921–24; the results of the Land Commission survey in 22 provinces during 1934–35; and selected materials from Chinese scholars written during the period 1920–36, compiled and translated by the Institute of Pacific Relations.[3] These classic studies of Chinese rural economy reveal the following broad characteristics:

1 Landownership was concentrated, varying in degree from one region to another, but the scale of large private farms was far below that of today's Latin American countries, Pakistan or even that of Egypt prior to 1952 partial land reform. The size of large Chinese farms ranged from 20 to 1,380 hectares, with the average size 2,000 mu (335 acres or 139 hectares). The number of cultivators of dominant small and fragmented holdings represented about 70 per cent of total landholders, yet the actual area of such holdings represented only 20–25 per cent of the total area of landholdings. Each owned and/or rented a farm of 15 mu (2.5 acres) on average, many of which were fragmented into six to nine scattered plots. According to a comprehensive farm management survey of 4,312 farms in Kashing during 1935, 59 per cent of the farms had no working animal stock (Institute of Pacific Relations 1939: Tables 12 and 13). In four provinces for which data on size distribution are available, the Gini coefficients are high, ranging from 0.540 to 0.735 (Table 5.2).

2 The scale of tenancy including sharecropping ranged from 30 to 50 per cent. Rental levels were high, ranging from 50 to 70 per cent of the harvest. Payment was made as a cash advance, thus assuring the landlord's income in case of crop failure.

3 The extent of pure landlessness was not high, with the proportion of landless to total agricultural households ranging between 20 and 30 per cent (as estimated by the studies cited). As the average number of working days per

Table 5.1 Estimated redistributive scope of state-administered land reform in 22 developing countries (excluding settlement schemes), 1915–94

Countries in descending order of beneficiaries' scale and years of reform acts	Beneficiary households as a percentage of total agricultural households	Redistributed land as a percentage of total agricultural land	Size ratio of land ownership ceiling to beneficiaries' unit sizes
China (1949–56)	About 90	80[a]	No ceiling
South Korea (1945, 1950)	75–77	65	3 ha : 1–2 ha (3:1)
Cuba (1959–65)	60	60	67 ha : 35 ha (2:1)
Ethiopia (1975, 1979)	57	76[b]	10 ha : 13 ha (1:1.3)
Iraq (1958, 1971)	56	60	Varies according to land quality
Mexico (1915, 1934, 1940, 1971)	About 55	42	100 ha irrigated and 300 ha rainfed ceiling and 2–5 ha irrigated units (28:1)
Tunisia (1956, 1957, 1958, 1964)	49	57[c]	Mostly recovered French-owned farms
Iran (1962, 1967, 1989)	45	34[d]	
Peru (1969, 1970)	40	38	Ceiling, irrig. 150 ha on coast and 55 ha in Sierra
Algeria (1962, 1971)	37	50[e]	About 40 ha : 11.5 ha. (3:1)
Yemen, South (1969, 1970)	25	47	8 ha : 1.2 ha irrigated (4:1)
Nicaragua (1979, 1984, 1986)	23	28	Ceiling 350 ha in Pacific zone and 4 ha (87:1)
Sri Lanka (1972, 1973)	23	12	25 ha : 13 ha irrigated (2:1)

El Salvador (1980)	23	22	
Syria (1958, 1963, 1980)	16	10[f]	120 ha : 1.5 ha. (24:1)
Egypt (1952, 1961)	14	10	Rainfed (7:11) irrigated (4:1)
Libya (1970–75)	12	13	40:1 irrigated
Chile (1967–73)	12	13[g]	Recovered former Italian farms 80 standardized ha and around 5 ha irrigated (16:1)
Philippines (1972, 1988, 1994)	8	10[h]	5 ha : 11 ha. corn and rice (3:1)
India (all, 1953–79)	4	3	Differs by states
Pakistan (1959, 1972)	3	4	65 ha : 4 ha (16:1).
Morocco (1956, 1963, 1973)	2	4	No ceiling, only recovered French-owned lands

Sources: China, South Korea, Cuba, Iraq, Mexico, Egypt, Sri Lanka, India and Pakistan: El-Ghonemy (1990a: Ch. 6 and 7, Table 7.1). Ethiopia: Abate and Kiros (1983: 160–76). Chile: Castilio and Lehman (1983: 249–68). Nicaragua: Baumeister (1994: 223). Peru: Kay (1983: 206–17). Chile and Peru are in Ghose (ed.) (1983). El Salvador: El-Ghonemy (ed.) (1984a: 20–1). Algeria, Tunisia and Libya: El-Ghonemy (1993). Iran: El-Ghonemy (1998: 157–9). Yemen, South: calculated from FAO (1984). Syria: estimated from FAO (1984, 1991). Philippines: El-Ghonemy (1990b: 269–72), Government of the Philippines (1990: 18) and Sentra (1997:15–31).

Notes
a After deducting areas of state farms.
b Area of peasant associations and including producers' cooperatives.
c Includes the individualized *habous* on private *waqf* land.
d Includes the area reallocated by the Council of Determination in March 1989, which was occupied by peasants after the owners fled the country.
e Includes 2.6 million hectares of recovered French-owned farms (autogestion socialist sector).
f The area does not include 911,201 hectares expropriated but not redistributed up to 1990.
g These estimated percentages of beneficiaries and land rise to 18 and 36 per cent, respectively, when all *asentados* (potential beneficiaries) were included (see Barraclough and Alfonso 1972: 16).
h After the deferment of the distribution of 0.3 million hectares to the year 2005, the restitution of nearly 80,000 hectares to original owners and the exemptions made in President Ramos's Decree RA7881 of 1994.

Table 5.2 Gini coefficient of pre-reform land concentration in four Chinese regions, 1929–36

District and province	Year	Gini coefficient
Wusih, near Shanghai (1)	1929	0.666
Chitung, northern Kiangsu (2)	1933	0.735
Chekiang, south China (3)	1936	0.674
Henan, north China (3)	1936	0.540

Sources: 1, Wong Ying-Seng, Chien Tsen-jui and others (1932) 'The Land Distribution and the Future of Capital', unpublished MSc thesis, 1932; 2, survey by the National Rehabilitation Commission (1933) Chang *1-pu Land Distribution and Tenancy in Kiangsu. Chung-Kuo Nang Ts'un* 1935, Shanghai; 3, Sun Shao-Tsun (1936) 'The Land Problem of Modern China', *Education* November.

Note
The Gini coefficient is calculated from data given in Institute of Pacific Relations (1939: Tables 1.2, 1.3 and 1.4).

year was only 130, average annual earnings for workers were correspondingly low.
4 Among peasants, heavy indebtedness and land mortgages prevailed. Furthermore, peasants were heavily taxed, both formally and through informal levies in the form of land taxation and unpaid military service. Local officials, landlords and grain dealers acting as middlemen abused the tax collection.
5 The consumption of poppy-opium damaged the health and economic position of the peasants. Approximately 40 per cent of the total adult male population were addicted to the narcotic. Addiction was particularly high in Szechuan, Fow-Chou and Yunnan.

The agrarian system features outlined in these five points had serious consequences for the performance of agriculture and, accordingly, for rural poverty. Investment in labour-using and yield-increasing technology was negligible. This type of investment can hardly be expected in conditions in which 10 per cent of total landholders, as absentee owners and tenants, controlled 70 per cent of total cultivated land. Nor could insecure and indebted tenants or other poor peasants afford technological change. The proportion of irrigated land remained at a low 16 per cent around 1947. Yields of the main food and non-food crops were also low (Table 5.3). According to Ramon Myers, the cropping index rose slightly, but scarcely any new technical advances were introduced (Myers 1982: 43).

To appreciate the fundamental transformation brought about by the 1948 and 1978 reforms, we note that low productivity was not the only feature of pre-1949 Chinese rural society. Illiteracy, particularly among women, was high at 80 per cent, as was the prevalence of ill-health (widespread tuberculosis, malaria, cholera, smallpox and leprosy). Infant mortality was high at nearly 200 per 1,000, while average life expectancy, at 29 years, was extremely low. Scattered data suggest a rough estimate of absolute poverty of 60–65 per cent in the pre-1948 rural areas.

According to a Chinese scholar who analysed the Nanking University's field investigation in rural areas: 'While the general phenomenon among the rural rich is a trinity of landlord, merchant and moneylender, that among the rural poor is another trinity of poor tenant, hired farm labourer and coolies.' Thus, in 1948–49, Mao Tse-Tung rapidly instituted a series of land reforms designed to tap the productive power and latent abilities of about 250 million poor peasants and farmworkers.

The agrarian economy transformation 1948–78

The role of land reform in enhancing the process of rural development in China, the most heavily populated country in the world, cannot be underestimated. The programme held broad objectives and worked through several stages of implementation, beginning in 1948 amidst the devastation of the war with Japan, which began in 1937. Led by a large-scale land reform programme, the chief elements of the post-1948 land reform-based process of rural development are outlined in broad terms as follows:

1 Elimination of the power of landlords, moneylenders and traders in rural areas. To do so, land was expropriated without payment of compensation. It was then redistributed to the mass of peasants, tenants and landless workers on an egalitarian basis, thus ensuring greater equality.
2 According priority to agriculture as the foundation of national development. Emphasis was placed on the sustenance of high rates of growth in the production of food grains, based on labour-intensive technology (guided by the National Programme for Agricultural Development formulated in 1955). Self-sufficiency in supplying food grains at local and national levels was a chief goal.
3 Mobilization of the agricultural labour force. This required enhancing their abilities and converting their productive capacity into capital. Harnessing this enormous power for rural development was central to the formation of cooperatives and communes, which encouraged self-reliance in development and decentralization of decision-making to the local level.
4 Establishing small-scale, labour-intensive industries, spatially scattered in rural areas. Less developed areas were given priority for such industries, thus evening out inter-regional imbalances in incomes.
5 Balancing the demand for food with its supply by restricting both population growth and migration of rural people between and within regions.
6 Finally, at the heart of the process for transformation has been the development of human resources. As shown in items 2–10 and 15 in Table 5.3, improvements aimed at: universal literacy; accessible health and sanitation services, combining modern medicine with traditional methods (including barefoot doctors); equality between men and women; and motivating the peasants towards hard work.

The point to bear in mind is that the elements outlined above were rooted

Table 5.3 Selected indicators of agrarian and human capability changes in China, 1930–85[a]

	1930–40 average	1952	1960	1979	1980	1985
1 Income per capita agricultural population		43.1 (1957)		62.8 (1976)	85.9	
Total income (yuan) at constant 1957 prices		73.0 (1957)		113.0 (1976)	170.0	
2 Average food grain consumption (kg per capita)		197.5	163.5	188.0	212.5	
3 Daily caloric supply per capita	1,993 (1933)		1,942 (1961–63)	2,222	2,526	2,602
Daily calorie supply as a percentage of requirement	90		82	97	107	119
4 Infant mortality per 1,000	200		90	68	39	35
5 Life expectancy at birth (years)	29	36 (1952)	51	63 (1975)	67	69
6 Adult illiteracy rate (%)	78	60		34	30	
7 Index of state investment in agriculture (1952 = 100)		100	264	479 (1965)	993 (1979)	
8 Irrigated land as a percentage of arable land		19.7	29	37	44	
9 Fertilizer use (kg per hectare arable land)	2 (1949)	2	5	41	150	192
10 Per capita food grain production (kg)		285	215	291	326	350
11 Crop yield of cultivated land (tonne/hectare)			1965	1975		
Wheat	1.0		0.9	1.4	2.0	2.9
Rice (paddy)	2.5		2.7	3.0	4.2	5.2
Cotton	0.5		0.5	1.5	1.6	2.2

		1960–70 average	1971–80 average	1981–85 average
12 Annual rate of population growth (%)	2.0	2.3	1.8	1.2
13 Agricultural labour force rate of growth (%)		1.6	1.9	1.5
14 Agricultural production rate of growth (%)		6.2	3.8	6.5
15 Per capita agricultural labour productivity rate of growth (%)		4.6	1.1	5.0

Sources: In the order of indicators as numbered: 1 and 2, Lardy (1983: Table 4.6 for item 1, and Table 4.3 for item 2). 3, for 1933 Lardy (1983: Table 4.1) (Weins' estimate). For the rest, Fourth World Food Survey, Rome: FAO, 1977. Appendix C, and *FAO Production Yearbook*, Vol. 3 and Vol. 39. 4 and 5, for period 1930–40, Perkins and Yusuf (1984: 133–7). The rest from World Bank *Development Indicators*. 6, as in (5) except Perkins and Yusuf (1984: 171–2). 7, calculated from Perkins and Yusuf (1984: Table 2.6). 8, for the year 1952, Yeh, K.C. (1973) in Yuan-lium (ed.) *China, A Handbook*, Newton Abbot, Table 20.2. The rest from *Country Tables*, Rome: FAO. 9, for 1949 and 1952, Aziz, Sartaj (1978: Table 3.2). The rest from *Country Tables*, Rome: FAO. 10, Lardy (1983: Table 4.2) except 1985, calculated from data in FAO *Country Tables* 1987. 11, *FAO Production Yearbook*, several volumes. 12, World Bank, *World Development Report: Indicators*, several issues. 13 and 14, Calculated by the present author from data on total agricultural production in physical terms and average annual rates of growth of agricultural labour force. 15, FAO *Country Tables*, 1987.

Notes

a Post-1985 data are not included, leaving the changes to Chapter 8 in the section discussing the effects of economic reforms.
Yuan exchange rate was US$1 = 2.50 yuan in 1950, 1.54 yuan in 1979 and 1.71 yuan in 1981.
Collective income, in cash or kind, distributed to production team members out of the net income realized by the team.

in Chinese ideological thinking and gradual adjustments over a period of three decades from 1949 to 1978 through a pragmatic approach relevant to Chinese traditions. The initial stage was the redistribution of land by 1952, based on individual ownership of land, and producers' cooperatives provided the necessary means of production. According to Kenneth Walker (1965: 5) and Sartaj Aziz (1978: 10), about 47 million hectares of cultivable land was equally distributed 'among 300 million landless and land-poor peasants each receiving an average of 0.15 hectares (0.4 acres)'. An additional 4.2 million hectares (4.3 per cent) was converted into state farms.

An indication of 'a return to capitalism' in production relations, however, moved the country's leadership to take further steps towards collectivization. It was claimed that some post-land reform owners were 'uncooperative', and others sold their land to other peasants and worked as wage-based labourers. Accordingly, all individual holdings were collectivized and their ownership rights changed to collective cooperatives. This was accomplished *nationwide* with dramatic speed, but at the expense of a short-term decline in grain and sugar cane output (1958–62). Strong technical support to the cooperatives was provided by the government and, in 1958, local administrative units (*hsiangs*) were amalgamated with the cooperatives to form communes. Small private plots for the use of peasants were individually allocated (as kitchen gardens and for raising chickens and pigs for family consumption or sale). Peasants were rewarded for their farm labour and for commune-run non-farming enterprises on a work points basis.

Post-1978 changes in production incentives and rewards

With some minor variations between regions, the system described above continued until 1977, 1 year after Mao's death. Then, ideological struggles, too complex to be outlined here, erupted among the new leadership (White 1983; Khan and Lee 1983). Since 1978, institutional changes have been introduced to improve the peasants' production incentives and incomes. These changes may be classified into three subsets of institutional arrangements.

First came arrangements to promote the individual producer's material incentives and to motivate his or her production and exchange activities. Their aim was to shift primary consideration from production alone to increased consumption and trade services. No longer was household income determined entirely by the distribution of collective income at the commune/team level (based on working points earned). Instead, incomes were gradually to be linked with the household's own production and the volume of sale of marketed surplus in the free market. Under the new institutional arrangements, an autonomous entity called a production team (consisting of 30–40 households) delegated production responsibilities to either an individual or a group of households. The production team represented the state as the owner of communal land and major capital equipment. We should keep in mind that post-1978 changes maintained the state's property right of land, which was not privatized.

The accounts given by Khan and Lee (1983), Perkins and Yusuf (1984) and

Griffin (1984) show the prevalence of one arrangement. An individual household is contracted for a specified period of time to have the right to use a plot of land, equally allocated to men and women on either a per capita or a per worker (adult) basis (average about 1 acre per family). The household has the legal obligation to pay a share in land taxation as well as in welfare funds for health and education services. He or she must deliver a fixed quota from the total produce at a price fixed by the state. The balance of goods may be retained for sale at the 'free' market price. Income differentiation may arise from variations in the quality of land and the portion of non-collective income relative to total income accrued to each household. In a sense, such contracting arrangements resemble tenancy in many developing countries that have private property market economies

The second new institutional arrangement was the revival and expansion of households' 'personal' plots of land, which lie outside communally cultivated and managed land. These plots are not subject to either the state-controlled cropping pattern or to the delivery of a prescribed quota of produce. This new incentive has raised the portion of personal cash income from plots, and has provided households with increased security of food intake.

Finally, in order to match food demand to supply, post-1978 policies strictly control births. The preferred number of children for each married couple is one, and births are restricted to two. Rural migration to urban areas is also restricted. These restrictive measures were intended to: (i) realize an ambitious plan to raise per capita income from its average US$300 in 1981 to US$1,000 by the year 2000; (ii) limit government expenditure on public services and food subsidies; and (iii) reduce the rate of growth in food grain consumption.[4]

Within an emerging market mechanism, these institutional arrangements provide material incentives to strengthen private consumption, and to increase household savings and capital accumulation.

Redistributive consequences

Unique characteristics have emerged from this continuous but gradual transformation of the agrarian economy, which began with a substantial redistribution of material assets and incomes. The first is that the benefits were achieved by relying on Chinese resources and by pursuing approaches to social change relevant to both the country's conditions of poverty and a chosen ideological path. They tend to be supported by the peasantry's inspired traditional Confucian values of discipline, obedience, patience and cooperation. The second characteristic is the speed and high implementation capability with which institutional, technical and social transformations of the agrarian economy were realized. The third is the magnitude of the reduction in poverty within rural areas where the confined population was not permitted to migrate to the cities. The fourth is that, despite the substantial public expenditure on the improvement in quality of life, the low-income Chinese economy was able to achieve a high share of gross domestic savings and investment in GDP. In 1965 and 1985, the share of the former amounted to 25 per cent and 84 per cent, and the latter was 25 per cent and 38 per cent respectively. Gross domestic investment grew during the period at an average annual rate of 13 per cent, nearly four times the rate of other low-income countries (World Bank 1987, *World Development Report*: Tables 4 and 5, Indicators).

Poverty reduction

Perhaps the most remarkable change in the countryside since the start of the Chinese complete land reform in 1948 has been the fast reduction in income-based absolute poverty from roughly 60 per cent before the reform to a range of approximately 6–11 per cent during the period 1979–81. At the same time, there has been a sustained reduction in the number of poor persons from about 240 million to approximately 50–80 million over this period. According to official data, the rural poor in 1981 were concentrated in 87 counties in four provinces.[5] This achievement is attributable to persistent commitment on the part of the country's leadership to provide the mass peasantry with accessible opportunities for secured and equal access to land, a guaranteed minimum level of food grain consumption (150 kg of wheat or 200 kg of rice per capita per year) and rapid and significant reductions in illiteracy and infant mortality. All these have contributed to a substantial 130 per cent rise in life expectancy at birth from about 30 years in the 1930s to nearly 70 years in 1985 (see Table 5.3).

Obviously, the complete land reform had the greatest effect in rapidly reducing gross inequalities in land distribution (and hence income), from a Gini index of land concentration of 0.7 in the 1930s to very equal distribution of land. The Gini index of income distribution in rural areas was 0. 211 in the 1970s according to Perkins and Yusuf (1984). This equality in the distribution of productive assets has enabled widespread benefits from the realized agricultural growth.[6] However, there have been, and will continue to be, marked regional differences in per capita income and consumption. An influential source is the marked variation in natural endowment: the amount and quality of agricultural land and climatic conditions. The threefold expansion in the irrigated area since 1949 from 16 to 45 per cent in 1985 is a crucial contributing factor.

From the experience gained during the initial period 1978–83, Keith Griffin and his five associates concluded from their empirical field studies:

> We have seen that, in practice, rural China remains a remarkably equal society and no statistically reliable evidence exists to show that the degree of equality has diminished since the post-1978 reforms were introduced. Those who believe the contrary have had to rely on anecdotal evidence . . .. If in fact, income inequality and social stratification do become serious problems in the years ahead, the explanation probably will lie with changes in the relations of production.

> Griffin (1986: 310)

After examining evidence collected during 1982 and 1983 about ownership and the use of means of production, they added: 'As we have seen, there is *no evidence* yet of increased inequality [my emphasis]. Hence, it is best, perhaps, to continue to regard the current period as one of experimentation, albeit on a national scale' (Griffin 1986: 315).

To conclude this long dynamic process of adjusting the institution of property rights and the structure of production incentives and rewards in China's gigantic rural sector, we must realize that the agrarian economy has been characterized by social ownership of the major productive assets. In this process, planning and the market mechanism are considered complementary and *not* as alternatives. But will it work? Can the Chinese economy realize greater economic growth, private consumption, savings and investment and, at the same time, maintain the high degree of equality in income distribution achieved up to 1978? We shall find out later in Chapter 7.

South Korea

Like China, rural development in South Korea was based on egalitarian distribution of assets and income at an early stage of national development. In 1945, South Korea instituted complete land reform with centralized planning and labour-intensive agriculture that would employ abundant labour and scarce land. But unlike China, policy choice in South Korea was *private property* within a market economy, which is fundamentally capitalist but regulated by the state. Whereas China had based its land reform policy on nationalizing land (with distribution free of payment), South Korea chose contractual transactions with payment of compensation by beneficiaries to affected landlords. Finally, South Korea's reform was initially induced by an external agent, the United States Liberation Forces, after the defeat of Japan in the Second World War. Nevertheless, the experience of South Korea in rural development offers innovative ideas.

The pre-1945 agrarian system

The discussion that follows of the pre-1945 agrarian system draws upon the following sources: Lee (1936); The United States Agency for International Development (1970); Sung-huan Ban (1980); Keidel (1981); my own discussion with Dr Clyde Mitchell, who was the American economist in the US Administration of South Korea (1947–50); and my interview with Dr Hyuk Pak in 1968, when he was in charge of his own country's land economics research.

Two major surveys recorded the land tenure systems and production relations in agriculture for the decades immediately preceding 1945. One was a cadastral survey carried out by the Japanese colonial administration during the period 1910–18 for land taxation purposes and for establishing Japanese ownership of a large portion of agricultural land. The other survey identified the nationality of landowners (Japanese, Korean, Chinese and other foreigners) in 1927. Together, these surveys were useful bases for formulating and implementing the land reforms of 1945 and 1950. They illustrated the great inequality of existing landownership. According to the surveys, Japanese settlers representing only 1.3 per cent of total landowners possessed almost 55 per cent of total South Korean irrigated land in 1930, with an average ownership of 100 cho (240 acres) each (Lee 1936: 149). Compared with the already high Gini index of 0.624 for all Korea, this index for irrigated areas was even higher at 0.823 (Table 5.4).

In addition to the high degree of land concentration, absentee ownership was widespread. Accordingly, the extent of tenancy in the provinces of South Korea rose quickly from 40 per cent in 1920 to 56 per cent of total landholders in 1938, falling to 49 per cent in 1945. Based on his field survey of a sample of 1,249 farming units in 1931, Hoon Lee reported that most tenants were burdened with indebtedness through high rent payment (50–70 per cent of harvest) and lacked access to institutional credit. Forced to rely heavily on landlords and moneylenders, the tenants were charged interest rates ranging from 40 to 70 per cent per year. Landless agricultural workers accounted for 30 per cent of the total agricultural households in all Korea. Based on his calculations of the gross and net income of agricultural households by tenure status, Keidel roughly estimated the poverty prevalence (landless and small owners and tenants) at 60 per cent of total agricultural households in South Korea in 1925 (Keidel 1981: 45, Table III-9). This state of poverty was dramatically described by Hoon Lee as follows:

> They survived (outside the working seasons) by eating millet bran, legume pods, tree bracken, grass roots. They live because they cannot die . . .. When sickness and disease befall them, their fate is doomed. Health services and administration of hygiene in these rural sections are far behind the times.
>
> Lee (1936: 171, 172)

Table 5.4 Distribution of land ownership – all Korea, 1927 and 1930

Size classes in cho	1927 (all ownerships)		1930 (irrigated ownership)	
	No. of owners (%)	Area (%)	Number (%)	Area (%)
Less than 0.1	18.00	1.30	27.0	0.7
0.1–0.5	32.00	6.40		
0.5–1	19.40	10.10	16.0	1.2
1–5	26.00	42.70		
5–10	3.00	15.27	45.5	15.5
10–20	0.90	8.95		
20–50	0.54	9.63	8.5	18.3
50–100	0.14	2.90	1.5	10.1
100 and over	0.02	2.75	1.5	54.2
	100.00	100.00	100.00	100.00
Gini coefficient	0.646		0.823	

Source: Calculated from data given in Lee (1936: Table 61 for 1927, data on p. 149 for 1930).

Note
One cho is a little less than 1 hectare and equals 2.4 acres.

The 1945–50 complete land reforms

The United States Military Forces administered South Korea for 3 years from 1945 to 1948. But they did not enforce redistribution of privately owned Korean land as they did in Japan. Instead, the US administration chose to: substantially reduce rents; secure tenancy rights; take over the 324,464 hectares formerly owned by Japanese settlers for redistribution among actual tillers; and play a key role in the land reforms of 1949 and 1950 following the establishment of the Republic of Korea in 1948. The 1949 Land Reform law and its 1950 amendment were implemented amid the turmoil of the 1950–53 war between North and South Korea, which devastated agriculture and took 1.3 million lives. The following fundamental changes in the institution of property rights in land were effectively introduced during the period 1945–53:

1 The outright transfer of income in real terms from landlords to tenants by the substantial reduction in rent from an average 60 per cent of the harvest to a maximum of 33 per cent. This was accompanied by the provision of a higher degree of security of farmland tenancy.
2 The direct sale of 573,000 hectares (28 per cent of the total cultivated area) from Korean landlords to their tenants.
3 The redistribution of 245,554 hectares, formerly held by Japanese owners, to the tenants who were cultivating the land.
4 The Korean government's purchase of 332,000 hectares from landlords whose landholdings exceeded the prescribed size ceiling of 3 chungbo (a little less than 3 hectares or 7 acres) per owner. Former landlords were encouraged to invest the value of compensation payments in industry. The new owners, on the other hand, paid the government the full value in addition to land taxes. Both transactions were efficiently implemented within 5 years.
5 A programme for investment in agriculture began in 1954 and, with substantial aid from the United States, the Agricultural Credit Bank was established.
6 Distribution of the *total area* amounting to 1,150,554 hectares (listed in 2, 3 and 4 above) in plots averaging 0.9 hectares.

The dynamics of securing accessible opportunities, 1955–85

Having briefly characterized the large-scale land redistribution, it had four primary consequences. The first was income transfer resulting from combined rent reduction (most in kind) and freeing tenants from their accumulated debt. Tenancy was dramatically reduced from 49 per cent to about 4 per cent of total agricultural households (and, in fact, was 'officially' considered illegal by land reform legislation). Second, the number of new owner-operators (mostly former tenants and landless workers) reached nearly 75 per cent. Hired landless workers diminished from 30 per cent of total agricultural households to about 3 per cent who were gradually absorbed into non-farming activities. Third, with the very low ceiling prescribed for expropriation, about 65 per cent of the total area of cultivated land was redistributed. In addition, there was a rapid reduction in land concentration and a corresponding reduction in inequality in income distribution. Table 5.5 shows the sharp decline in the Gini coefficient of inequality for land distribution between 1945 and 1965 from 0.729 to 0.384.[7] Finally, we should not understate the intangible and unquantifiable improvement in the South Korean peasants' sense of dignity, self-respect and production incentives provided by individual landownership.

Having redistributed the scarce productive assets and removed the institutional barriers to participation in an egalitarian system of private property market economy, land reform set the stage for dynamic rural development. To gauge its interaction with the rest of the economy, it is important to consider changes since 1955 against a background of 35 years of Japanese colonial rule (1910–45), disruption in commercial and government services resulting in the 1945 partition and, finally, the war between North and South Korea from 1950 to 1953. Irma Adelman summed up this background:

> The Korean fortunes in the South in 1945 were rapidly eroded by the economic chaos caused by the partition; . . . the loss to the North of all heavy industry, major coal deposits, and almost all electric power generating capacity . . . and the flood of over 1.5 million refugees from the North . . . property damage resulting from the fighting has been estimated at US$2 billion. Agricultural output dropped by 27 percent between 1949 and 1952 and real GNP by 12 per cent. Prices rose by 600 per cent between 1949 and 1952.
>
> Adelman (1974: 281)

Agricultural output growth

There are problems in identifying the effects of the Korean land reform. For an intertemporal comparison before and after 1945, two problems arise. One is the change in the country's boundaries after 1945. The other refers to income comparability, and pertains to the choice of price index for deflation. For example, there are two sets of data on the annual rates of growth of agricultural production during the periods 1935–45 and 1945–53. Sung-huan Ban's estimate covers crop and livestock output using the 1934 price index. Keidel, on the other hand, included only crops and used 1970 prices for deflation to compare the values in constant prices (Keidel 1981: Table 111-2). Table 5.6 illustrates the different findings for annual rates of growth for the value of agricultural output in South Korea.

Despite these comparison problems, the data show an upward trend after the full implementation of land reform in the late 1950s. The growth rates reached an impressive level in the 1960s by any international standard. With the agricultural labour force growing at the low rate of 0.9 per cent during the period 1960–70, labour productivity grew at the rate of 3.6 per cent (a much higher rate than the average of all developing countries at 0.6 per cent). So too did the average annual rate of food production increase, at the rate of 4 per cent, again higher than the

Table 5.5 Stability in the size distribution of landholdings after land reform in South Korea, 1960–80

Size in hectares	Before land reform 1945		After land reform 1960		1965		1970		1980	
	N (%)	A (%)	N (%)	A (%)	N (%)	A (%)	N (%)	A (%)	N (%)	A (%)
Under 0.5	33.7	11.3	71.0	53.1	33.93	12.44	66.9	38.4	64.00	36.98
0.5–1.0	33.4	12.3			31.76	26.68			29.56	43.15
1.0–2.0	22.9	14.1	24.1	33.4	25.66	40.49	26.4	40.5	5.05	12.76
2.0–3.0	10.0	62.3	4.9	13.5	5.57	15.31	6.7	21.1	1.48	7.11
Over 3					1.17	5.08				
	100	100	100	100	100	100	100	100	100	100
[a]Gini index	0.729		0.388		0.384		0.314		0.303	

Sources: 1945, USAID (1970: Table 3); 1960, 1970 and 1980, Agricultural Census results, tabulated by the Statistics Division, FAO, Rome; 1965, Lee (1979: Table 2.14).

Notes

N, number of landholdings; A, areas of landholdings.

a The different size class intervals over the period 1960–80 affect this Gini index resulting from the use of fewer number of classes in 1960 and 1970 by the World Census of Agriculture for a uniform classification and the use of more class intervals in the surveys of 1945 and 1965.

Table 5.6 Different findings for annual rates of growth for the value of agricultural output in South Korea

	1930–39 (%)	1939–45 (%)	1945–53 (%)	1953–61 (%)	1961–69 (%)
Ban (1980)	2.9	–3.5	2.1	3.6	5.1
Keidel (1981)	–	0.0	0.9	4.0	4.5

average of all developing countries at 2.6 per cent (*North Korean Yearbook of Agriculture*, 1980, and FAO, 1987).

Crop yield per hectare, another indicator of the performance of the restructured agrarian system, is free of valuation problems. By 1965, the yield (tonnes per hectare) of paddy rice, barley and wheat had more than doubled, and the cotton yield grew to four times that of the average during the period 1933–40. Contributing to this notable rate of growth were three types of technical change: the sharp rise in the use of chemical fertilizer; the fast expansion in public investment in irrigation, land reclamation and soil conservation; and the high cropping intensity. Irrigated land area as a percentage of total arable land was more than doubled from 25 per cent in the 1940s to 57 per cent in 1985. Other institutions that effectively contributed to rapid rural development include newly established cooperatives, agricultural credit facilities, health centres and the innovative Saemaul Undong movement, which consolidated government services in villages and enlisted the effective participation of rural households in developing the social and production potentials of their own communities.[8] The egalitarian base of the rural economy and the introduction of incentives have undoubtedly been primary ingredients in the process of sustained agricultural growth and the consequential alleviation of rural poverty.

Equitable income distribution

Average farm income per household increased by 51.4 per cent in real terms between 1963 and 1975 (an annual growth rate of 4.3 per cent).[9] Rises in both per capita rural population and per agricultural household income have been reinforced by spectacular expansion of educational opportunities and rapid outmigration of the rural population at an annual rate of 1.5 per cent between 1960 and 1970. The net rural migration to urban areas was estimated at 3.6 million, 65 per cent of whom were of working age, and 40 per cent of whom were female (Ban 1980: Table 137).[10] The allocation to education represented between 50 and 63 per cent of the central government expenditure on social services, and 2.5–3.5 per cent of GNP during the period 1961 to 1985 (*IMF Government Finance Statistics Yearbook*, various issues). In fact, agricultural population declined during the 1960s at the negative annual rate of 1.4 per cent. Rural outmigration and the substantial fall in the agricultural labour force have operated on the supply side of the rural labour market as a significant factor in raising the real wages and per capita incomes of those who remained on the land.

Several important lessons could be learnt from the successful experience of

South Korea. The first is that the high implementation capability for this large re-distributive programme within a stable political climate made these achievements possible. Government-induced greater investments for employment creation and favourable terms of trade for agriculture were combined in consecutive 5-year plans within a market economy. Second, in a private property market economy, government control of the reallocation of resources, along with preferential pricing and taxation policies, contributed to the stability of the pattern of income distribution in rural areas. At a Gini coefficient of 0.298 in 1963–65, the low degree of inequality was stabilized – with slight variation – and agricultural labour wages rose steadily in real terms (Lee 1979: 36, Table 2.7). This equality in rural areas is greater than the national Gini coefficient at 0.344 in 1965. Third, the beneficiaries were motivated to increase their saving ratios (savings out of average household income). Lee (1979: Table 2.12) indicates that the new owners of less than 1 hectare were able to raise their savings from 4 per cent of their annual income in 1963 to 16 per cent in 1973. These savings were encouraged by the government by their exemption from taxes under the small savings promotion scheme. Fourth, the rapid expansion of non-farm employment within rural areas has contributed to the stability and strength of the egalitarian rural economy.

The results of Korean agricultural censuses and Figure 5.1 indicate a generally stable pattern of landholdings between 1969 and 1980, despite a small variation.

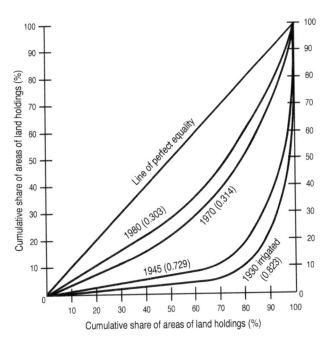

Figure 5.1 Lorenz curves for changes in the degree of inequality in land distribution in South Korea, 1930, 1945, 1970 and 1980. Prepared by the author from data in Tables 5.4 and 5.5. Figures in parentheses are Gini coefficients.

This variation is more noticeable between the size classes below 1 hectare than those of 1–2 hectares. The decline in the share of the former class is more likely to result from their sale and renting out of land in favour of migrating to urban centres and abroad.

Rural poverty reduction

The question now is: to what extent has the South Korean land reform-based rural development reduced the prevalence of absolute poverty in rural areas? Given the sustained rise in the average per capita income and wages in real terms, and given the political commitment for complementarity between market mechanisms and a government regulatory role to sustain an egalitarian rural economy with fast growth, a substantial reduction in rural poverty should follow. One way of identifying that reduction is to consider changes in the quality of life in rural areas. The 1979 nutritional survey of the Ministry of Health and Social Affairs shows that the average daily calorie intake per capita of the rural population was 2,707 or 22 per cent above the minimum requirement, and that only 3 per cent of the rural children under the age of 5 years were below the Korean standard for height-for-age. No deficiency in weight-for-age anthropometric measures was found (Dong Wang and Yang Boo 1984: Tables 61 and 63). These are marked improvements in food consumption and nutritional levels. This accomplishment was confirmed by the *Fifth World Food Survey* (FAO 1985a: Appendix lc), indicating a rise of 39 per cent in average daily calorie supply per person between 1961 and 1963; and between 1979 and 1981.

With regard to education, national illiteracy rates fell to 12 per cent among women and 4 per cent among men in 1983 (UNICEF 1984). According to the Ban study (1980: Table 135), similar progress was made in health. Infant mortality fell from 258 per 1,000 in 1945 to 42 per 1,000 in 1970 and to 34 per 1,000 in 1980. Accordingly, a most important achievement in human capabilities was realized; life expectancy at birth rose on average from 52 to 62 years for men, and from 56 to 71 years for women between 1960 and 1980. These indicators of the quality and quantity of life manifest the real meaning of development. They also point to a dramatic improvement in the characteristics of poverty.

Using available estimates of rural poverty referring to different points in time with different cut-off points and by different researchers is not entirely secure; yet, it is useful to consider them (Table 5.7). However approximate the estimates of the scale of rural poverty, they suggest a remarkably rapid reduction in both the proportion and the absolute number of rural poor in a private property market economy. We should not forget that, since 1950, the complete state-administered land reform programme has laid a foundation for the realization of rural development as defined in Chapter 2.

Table 5.7 Available estimates of rural poverty in South Korea

Reference year	Estimated incidence (%)	Approximate number of rural poor (in millions)	Source
1925	60.0	9	Keidel (1981) for provinces of South Korea
1965	40.9	4	Dong Wang and Yang Boo (1984)
1978	11.0	2	World Bank, Social Indicators Data Sheets, June 1984
1980	9.8	1.6	Dong Wang and Yang Boo (1984)

Iraq

While the news about Iraq's transitional political system and related civil conflicts has dominated the world public media since 2003, many readers might not know of its unique experience in creating an egalitarian agrarian system within a capitalist economy based on a complete land reform of 1958 and 1970. This case study is based on first-hand knowledge gained during my several visits in the 1960s and 1970s in my capacity as Chief of the Land Reform Division of the UN/FAO and from successive invitations by the government. In this case study, we focus on a country whose experience and economic structure are unlike those of China and South Korea. Iraq is endowed with rich natural resources: oil, abundant cultivable land, and water provided by adequate rainfall in the north (300–600 mm per year) and her two big rivers, the Tigris and the Euphrates. Furthermore, Iraq is not overpopulated (the density of agricultural population is only one-tenth that of China and South Korea), yet these resources were undeveloped. Prior to land reform, the Iraqi rural economy suffered from a high degree of land concentration, extensive land use, operating within a tribal system and exploitative tenancy arrangements. Agriculture, functioning at low productive capacity, was stagnant. Thus, amidst an abundance of cultivable land and oil revenues, a large section of Iraq's relatively small 3.9 million rural population in 1957 lived in poverty and lacked motivation. The land reform of 1958 was introduced by the military-led July 1958 revolution, which abolished landlordism and attempted to diminish the extreme powers held by tribal chiefs (sheikhs).

The pre-1958 agrarian system

To understand the impact of land reform on rural development since 1958, consider the pre-existing land tenure system. Until the defeat of the Ottoman empire in the First World War (1919), Iraq was under Ottoman rule, and its land property (*miri*) was until 1858 officially owned by the Sultan in Istanbul, who granted inherited rights of use to the occupiers against payment of tax. When the Ottoman land code was issued in 1858 to register individual land for full private ownership (*tapu*), most of it was registered as huge private properties of tribal chiefs, town notables and Kurdish Aghas (village heads in north-east Iraq) over the heads of the peasants. Angered by this injustice, peasants began fighting tribal heads, the land title registration procedures were halted and, consequently, the state owned all unregistered land (*miri sirf*). Later attempts (1910–32) by the Mandate of British administration of Iraq to re-establish registration failed. Sir Ernest Dawson, who was managing this work, wrote in 1932: 'today only a fraction of the cultivated land is somewhat uncertainly held on tapu tenure'. According to Saleh Haider (1944) and Doreen Warriner (1948), by 1943, only 17 per cent of agricultural land was registered as private property (*tapu*), and most of this was in the name of the politically influential tribal chiefs and city merchants, in addition to 5 per cent registered as religious endowment. The balance, 78 per cent, was officially the property of the state [made up of 22 per cent *lazma* land held communally by the tribes (*Dirah*) under lease from the state, and 56 per cent *miri sirf*, or pure property of the state). Charles Issawi gave this explanation: 'under the Mandate, the British introduced minor improvements, but did not attempt to alter the system for fear of antagonizing the landlords and tribal chiefs, on whose support they were dependent' (Issawi 1982: 147).

Following independence from British rule in 1932, the process of land accumulation by the sheikhs in shaping national policy was reinforced. Two laws issued in 1940 and 1945 for the sale of the state land, *miri sirf*, expanded the landed property of the sheikhs and city merchants, particularly in the southern provinces and the Sinjar region of Mosul province in the north (Ali 1955). Using increased proceeds from oil royalties in the early 1950s, attempts were made to redistribute a fraction of state land to landless agricultural households in seven land settlement schemes. In addition, the Development Board (*Majlis Al-Imar*) allocated 40 per cent of its development funds in 1951–58 to flood control irrigation, land reclamation and settlement schemes. While some of these schemes were successful and benefited settlers, other schemes failed through defective planning, land salination from expanding irrigation without drainage and the resulting abandonment of distributed units.[11] As an alternative to land reform, this slow and narrow approach was a short-sighted illusion, and manifests the policy-makers' intention to bypass the real issues in the agrarian system. It did, however, serve as a training ground for technocrats, many of whom, for the first time, came directly in touch with the realities of rural Iraq, recognizing the aspirations of poor peasants to own land.

The realities were quantified by the 1958 Census of Agriculture. Some 0.6 per cent of landholders held (ownership and lease from the state) 47 per cent of agricultural land in the size class over 1,000 hectares. In this category, five highly influential tribal sheikhs held 4 per cent of the total land, each of them owning (or holding under *lazma* tenure) more than 25,000 hectares. At the other extreme, 58,000 peasant households (mostly tenants) in the size group of less than 1 hectare represented 34.4 per cent of total landholders. Among them, they held only 0.3 per cent of the total landholdings area.

The 1958 and 1970 land reforms

The implementation of the 1958 land reforms in Iraq began with a vigorous process of expropriation of the large, privately owned and leased estates. Against the payment of compensation, those estates exceeding the prescribed limit of 2,000 donums (50 hectares) in rainfed areas of the north and 1,000 donums (250 hectares) of irrigated land in central and southern Iraq were expropriated for redistribution to the peasants (*fellaheen*). Lands belonging to the royal family and to those families considered as the 'enemies of the revolution' were confiscated without compensation payment. The total area affected was roughly estimated in 1964 at 8–10 million donums, or 2–2.5 million hectares.[12] (A donum or Mishara is equal to 0.62 acres or 0.25 hectares.)

Implementation of the reforms of 1958 suffered from several difficulties. Three factors emerged from my successive visits that offer some explanation.[13] The first is the political instability manifested in three *coup d'états* during the 1960s. Identified in Iraq as revolutions, each espoused a different ideological base for policy choice. Although strong commitment to land reform was maintained, the political conflicts and different promises confused the peasants (see Gabbay 1978). Conflicting questions of policy choice included whether to redistribute or to retain for the state the land in excess of the established size ceiling, and the functions of cooperative organizations. A second factor influencing the implementation of complete land reform in its initial phase was the insufficient capacity of the state institutions, which were unprepared for the speedy changes affecting two-thirds of the country's agricultural land and a major sector of total agricultural households. Furthermore, the 1958 law stipulated that this formidable task be completed within 5 years. Politics aside, practice proved that redistribution was difficult in the absence of undisputed records of land title registration, adequate numbers of trained staff and a network of institutional credit supply and marketing services to replace the functions of the ex-landlords. Most importantly, redistribution of land, particularly in the south, would be useless unless accompanied by secure access to irrigation water, treatment of soil salination and making arrangements with the former landowners (mostly tribal chiefs and their managers or *sirkals*) who owned the water pumps. The third factor, which I observed during my field study of 1964, that influenced implementation was the deeply rooted tribal affiliation of the peasants (*fellaheen*). Even in the face of the eroding power of the tribal chiefs, the affiliation of the peasants remained strong enough to their traditional chiefs to frustrate ongoing government efforts to organize agricultural cooperatives.

At the village level, a number of practical problems emerged from my field study during 1964 in the provinces of Amara and Hilla (renamed in 1971 as Maysan and Babylon respectively). From my discussions with tribal sheikhs, *sirkals* owning water pumps, heads of local government offices and a jurist (El Sayed Jawad Al Awady), I found that the following conditions existed:

1 Former holders of 3,000–5,000 donums under *miri lazma* for three decades were considered illegal landholders and were left with 150 donums (later raised to 300 donums), instead of the 1,000 donums that they expected to retain from the land reform law.
2 Whereas the sheikhs and *sirkals* had most of their assets expropriated and their official tribal power abolished, their influence remained intact because their production functions had never been replaced. This included supply of water from pumps, credit needs of beneficiaries and marketing their produce.
3 Lastly, there was an extreme shortage of staff whose time was spread thinly over the tasks of expropriation, redistribution, solving problems about land rights and the management of the expropriated large areas temporarily kept for eventual redistribution.

Despite these practical problems, the present author found in 1964 that, within 4 years, 80 per cent (about 7 million donums) of land subject to expropriation was already requisitioned by the state institutions. Out of this area, only one-fifth (1.5 million donums or 373,000 hectares) was redistributed to 45,000 peasant households. The remainder was kept under temporary administration by the newly established Ministry of Agrarian Reform. This large area was leased to the would-be owners numbering 250,000 families, each cultivating on average 25 donums.

By recognizing emerging problems and latent defects in the design of the 1958 land reform law, policy-makers made adjustments which culminated in the pro-socialism land reform law of May 1970. This second legislation pooled all former pieces of legislation and, importantly, lowered the size ceiling on private landownership in irrigated areas by 40 per cent. Collective farms were legally established by Article 38 according to the 'principles and rules of socialist co-operation'. In addition, the 1970 law allowed for variation in soil fertility, type of cropping and location of land in relation to market towns, and introduced flexibility about the minimum size of units to be redistributed in order to allow for a larger number of beneficiaries.

Despite the shaky database, the process of expropriation and redistribution was speeded up between 1975 and 1977, owing to the political settlement agreed between the central government and the leaders of the Kurdish provinces in the north. This ended the military confrontation and granted an autonomous administration in the northern provinces. Accordingly, 0.45 million hectares was redistributed to nearly 60,000 families, making the total number of new owners 218,000 households and the total redistributed area 1.92 million hectares by 1977. The 1970 land policy provided for the gradual allotment of the rented area under temporary administration as ownership units after the completion of drainage and irrigation work. The reclamation of state-owned land was also to be accelerated for establishing state farms. Three property rights institutions were adopted in the new system of land tenure: private family farms constituting most of the distributed area; collective cooperative farms; and state farms. We do not know the pre-1975 areas of collective cooperatives and state farms. What we do know is that they proved to be inefficient. In fact, by 1983, the response of the Iraqi peasants to collectives was negative so that their numbers fell rapidly from 77 in 1975, cultivating 120,830 hectares, to ten collective cooperatives in 1983, cultivating 7,543 hectares. According to Alwan (1985), many state farms were liquidated for inefficiency, and only 11 large farms, operating 170,000 hectares, were retained in 1983 for specializing in the production of cotton, sugar cane, sugar beet and sunflowers.

The impact of land reforms on agricultural growth, equity and poverty

Thus, after 12 years of uncertainty starting in 1958, the stage was set for land reform to lead a process of equitable rural development. Using plentiful oil revenue, the investment allocation for agriculture was substantially increased by 600 per cent between 1965 and 1978. For social services, including housing, health and free education and students' meals, the allocation rose from 115 to 1,081 million dinars (US$3.5 billion at the 1978 rate of exchange), a sharp rise of 840 per cent. As a relatively small contributor to GDP, agriculture's importance is in providing employment to a large section of Iraq's labour force (50 per cent in 1965 and 30 per cent in 1980). It is also responsible for providing most of the country's food supply and raw materials such as sugar cane, dairy products, cotton and sunflowers to domestic industry.

In sum, the scale of reforming the agrarian content of the economy was quite large. By 1985, an area of 2.4 million hectares was distributed to 262,000 agricultural households. If we add the areas of rented land (temporary administration), that of state-owned land under tenancy subject to distribution after completion of its development and the area under state farms, the total amounts to nearly 3 million hectares, representing about 60 per cent of the total area of arable land. If we add to the above number of land recipients those cultivating the balance of temporarily administered land expropriated and those tenants on the newly reclaimed state-owned land, estimated at 60,000 agricultural households, the total number of household beneficiaries reaches roughly 322,000 or nearly 56 per cent of total agricultural households in 1980.

AGRICULTURAL GROWTH AND PRODUCTIVITY

The insecure tenure arrangements and exploitative relationships of the pre-reform agrarian system imply very low production incentives or motivations and, therefore, the yields of the main crops were very low during the period 1948–58 (Tables 5.8 and 5.9). In 1964, during the present author's field study in Qalet Saleh, Nahr Sad, Abu Bishut in Amara province and Al-Shomaly in Hilla province, it became obvious that agriculture and the rural infrastructure needed everything to adjust from the production organization to the rapid change in the institutional framework. According to my study and despite the expensive water pumps installed (59 horsepower each), and canals and drains being constructed, I found that the cultivation of the fertile irrigated land was neither intensified nor diversified. The peasants in the land reform districts visited have continued growing the traditional

Table 5.8 Changes in the distribution of landholding in Iraq, 1952–82

Size of holding in hectares	1952[a] No. of holdings (%)	(Cum)	1958 No. of holdings (%)	(Cum)	1958 Area of holdings (%)	(Cum)	1971 No. of holdings (%)	(Cum)	1971 Area of holdings (%)	(Cum)	1982 Size of holding in hectares	1982 No. of holdings (%)	1982 Area of holdings (%)
<1	19.2	19.2	34.4	34.4	0.3	0.3	12.5	12.5	0.6	0.6	<2.5 ha	23.0	2.8
1–5	20.8	40.0	27.1	61.5	1.7	2.0	32.2	44.7	8.0	8.6			
5–10	12.8	52.8	11.2	72.7	2.2	4.2	23.6	68.3	16.9	25.5	2.5–30	72.1	66.7
10–20	15.2	68.0	10.4	83.1	4.2	8.4	20.3	88.6	27.7	53.2			
20–50	19.2	87.2	9.7	92.8	8.6	17.0	9.8	98.4	29.0	82.2	30–75	4.1	16.7
50–100	6.4	93.6	3.2	96.0	6.3	23.3	1.2	99.6	7.3	89.5			
100–150	1.8	95.4	1.0	97.0	3.4	26.7	0.2	99.8	3.0	92.5	75 ha and over	0.8	18.8
150–250	1.6	96.9	0.9	97.9	4.8	31.5	0.1	99.8	3.0	95.5			

	250–1,000	1,000 and over	Total percentage	Total	Gini coefficient
	2.3	0.7	100	125,045	0.902
	99.3	100.0			
	1.5	0.6	100	168,346	
	99.4	100.0			
	21.5	47.0	100	5,831,815	
	53.0	100.0			
	100.0	0.1	100	539,440	0.566
		100.0			
	100.0	4.5	100	5,156,027	
		100.0			
		100.0	100	682,864	0.394
			100	6,147,000	

Sources: Calculated from data given in 1952 Agricultural and Livestock Census results published in Statistical Abstract, 1956, Ministry of Economics, Government of Iraq. 1958 Census of Agriculture and Livestock results published in Statistical Abstract, 1960, Central Bureau of Statistics, Ministry of Planning, Republic of Iraq, Section VII, Table 90. In Arabic, the term 'ownership' was used, whereas in English, the results of the 1958 Census used 'holdings'. 1971 – Results of Census of Agriculture, Central Statistical Organization, Part 11. In our calculations, categories of 'holders without land' and 'over 2,000 donums' are excluded; the latter is mostly government and collectively managed as shown in Tables 3 and 4 of the Census results. 1982 data are from a report to the ruling Ba'ath Socialist Party held in Baghdad, January 1983 (p. 138) cited in Alwan (1985: Table 1). We do not know how the data were calculated in this report, which has fewer size classes than other years, making comparison difficult.

Notes

a Size distribution of holdings by area is not given in the 1952 Census of Agriculture for obvious political reasons.
Cum, cumulative share of number or area of each size class as shown by the Lorenz curves in Figure 5.2.

Table 5.9 Changes in agricultural income and rural quality of life in Iraq, 1948–80

Indicator	1948–52 average	1958	1961–65 average	1969–71 average	1974–76 average	1980
1 Agricultural per capita annual average income (dinars) at current prices	3.5ᵃ (1950)	26ᵇ (1960)	42ᵇ (1962)	69ᶜR (1972)	108ᶜR (1976)	1,58ᵇ
2 National (GDP)						
Average income per capita (dinars) at current prices	32 (1950)	51 (1956)	82 (1961)	138 (1975)	407	
At 1975 prices		110	134 (1961)	154	389 (1976)	
3 Daily average calorie supply per person	n.a.	1,856	2,012	2,678 R (1972)	2,839 R (1976)	2,840ᶜ
4 Item 3 as a percentage of requirement	n.a.	77	85	102	116	118
5 Infant mortality per 1,000	About 350 R	140 (1960)	121	100	111 R	73
6 Life expectancy at birth (years)						
Male	47			53		57
Female	50			56		61
7 Illiteracy rate as a percentage of adults	n.a.	84 R	82	66	n.a.	32 R
8 No. of water pumps for irrigation per 1,000 hectares of arable land	n.a.	1.1	1.3	1.4	1.8	n.a.
9 Fertilizer use (kg per hectare)	<1	1	1	3	6	17
10 Institutional agricultural credit (dinars per hectare of arable land)	n.a.	0.98	11.12	409.76	490.71	n.a.

11 Yield (tonne per hectare)

Wheat	0.48	0.4	0.53	0.88	0.77	0.72
Barley	0.77	0.5	0.75	1.19	0.93	0.72
Rice (paddy)	1.16	1.12	1.42	1.95	2.73	2.75
Cotton (seed)	0.27	0.46	0.74	1.39	1.33	0.87
Cotton (lint)	0.14	0.18	0.24	0.40		0.57

	1953–61	1961–70	1971–80
12 Agricultural GDP average annual growth rate	1.5	5.7	–1.5 (1970–77)

13 Food production index

1952–56	1958	1961	1968	1969–71	1973	1974/5	1979/80
100	106	123	167	119	104	94	150

Sources: According to row numbers:

1 Letter 'a' refers to tenants and sharecroppers only (see text), 'b' refers to per capita agricultural GDP calculated from *FAO Country Tables*, and 'c' refers to per capita expenditure in rural areas taken from Bakir (1979: Table C21).

2 Current prices: 1950 and 1956 from Fenelon (1970: Table 4). Other years are from World Bank Development Indicators. Constant at 1976 prices: Central Statistical Organization cited in Bakir (1979: Table C30).

3 and 4 FAO *Food Balance Sheets* except 1972 and 1976 from Bakir (1979: Tables C12 and C13).

5 For 1949–52, data taken from Adams (1958). For other years from UNICEF *Statistics on Children*, May 1984.

6 *World Development Report* – Development Indicators, World Bank, several issues.

7 Data for rural areas from Bakir (1979). The rest from UNICEF (1984). Illiteracy rates in rural areas in 1970 were 68 among male and 96 among female adults.

8 and 10 Calculated from the results of the 1958 Census of Agriculture and from several issues of the Iraqi Ministry of Planning *Statistical Abstract and Annual Abstract of Statistics*.

9 *Country Tables* 1987, Rome: FAO.

11, 12 and 14 *Production Yearbook*, Rome: FAO; data for 1948–52 and 1961–65 are calculated from Vol. 24, 1970; 1969–71 from Vol. 31, 1977; 1974–76 from Vol. 30, 1976; and data for 1980 from Vol. 34, 1980.

13 Data for the period 1953–61 are from Hasseeb (1964). The rest are taken from The World Bank *Development Indicators*.

Note

Data are until 1980 when the war with Iran started. R refers to rural population or rural areas. One dinar = US$3.2 up to 1980. Figures in parentheses refer to corresponding year.

low-yield barley in two-thirds of their allotted holding area, and only one-third of the area is given over to high-yielding varieties of rice, which it was intended to expand. Nor were cropping patterns diversified as one would expect after the construction of the very costly irrigation schemes.

Discussing these economic considerations with the authorities in Baghdad made it quite clear that the question of realizing an economic return on the investment of oil money in modernizing agriculture and raising productivity was of little concern. Real output in agriculture did not seem to match the heavy public investment in irrigation and drainage. Furthermore, the prolonged drought of 1973–75 severely damaged the wheat and barley harvests, particularly in the northern region. In addition, the sluggish growth in rates of food grain and meat production could not keep pace with the faster rates of annual population growth from the already high rate of 3.2–3.6 per cent between the 1960s and 1970s. It is true that, with its plentiful oil revenues, Iraq could afford to make up for the balance of increasing food imports. But this is inconsistent with the overriding proclaimed national plan objective of food security and food self-sufficiency.

EQUITY AND POVERTY

To what extent have land reforms since 1958 shaped the income distribution in rural areas and affected nutritional levels or reduced the prevalence of poverty? The intangible social gain is the liberation of the peasants from their dependency for survival on the arbitrary powers of the tribal sheikhs and the *sirkals* who had been classic institutional monopolists.[14] The reforms have nearly eliminated such economic control.

A major effect has been the substantial and rapid reduction in the degree of inequality. Table 5.8 indicates that the share of land held by farmers holding less than 20 hectares each rose sharply from less than 8.4 per cent in 1958 to almost 53.2 per cent of total agricultural land in 1971. In addition, farmers in the size class of less than 30 hectares in 1982 represented 95 per cent of the total number of landholders and held 69.5 per cent of total land. Another significant effect of land reform is the corresponding sharp decline in the share of the area of holdings over 250 hectares from 68.5 per cent in 1958 to 4.5 per cent in 1971. This change towards an egalitarian agrarian system is clear in the reduced Gini coefficient of land concentration from the highly unequal index of 0.902 in 1958 to 0.394 in 1982 (Figure 5.2).

With regard to living conditions and income distribution, in 1951, Fenelon, the Director of Statistics in the Iraqi Ministry of Economy, calculated the national average per capita income at 31 dinars (Fenelon 1956: Table 4). For the hired agricultural workers, annual earnings were also low, as each worked about 100 days a year and earned about 150–250 filses a day plus, in many cases, a meal and tea.[15] Infant mortality was estimated at 350 per 1,000 in the early 1950s. On average, calorie intake per person per day was 77 per cent of calorie requirements. From scattered information (and in the absence of further data), we estimate that at least 70 per cent of the rural population lived in such poverty conditions. In his analysis

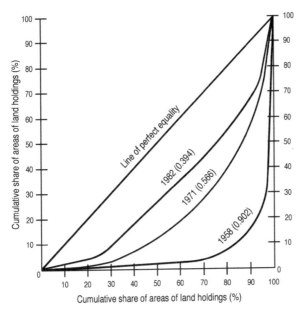

Figure 5.2 Lorenz curves for changes in the size distribution of landholdings in Iraq, 1958, 1971 and 1982. Prepared by the author from data in Table 5.8. Figures in parentheses are Gini coefficients.

of the results of the household expenditure surveys by localities in 1971, 1976 and 1979, Mohammed Bakir (1979) provides quantitative estimates of changes in the level of living standards and in the distribution of per capita consumption in rural areas. We need to keep in mind that *expenditure* distribution is usually less unequal than *income* distribution as the latter includes savings, the average rates of which are likely to rise as income rises.

Bakir (1984) calculated the Gini coefficient for the distribution of household expenditure in 1956 at the national level (rural and urban) at 0.630, which dropped sharply to 0.241 in 1972 and to 0.220 in 1979. For *rural* areas, the index shows a greater equality: 0.150 in 1971 and 0.173 in 1979. These are post-land reform very low inequality values, particularly compared with those in China and South Korea, requiring caution. Nevertheless, the substantial and rapid reduction in the degree of inequality after land reform is a remarkable achievement. At the same time, the fast rural outmigration has contributed to the increased per capita income/consumption of those who remained on the land.

But to what extent has land reform, combined with the effects of the rapidly growing oil-based economy of Iraq, reduced the prevalence of rural poverty? Unfortunately, neither the government statistical office nor the Iraqi scholars concerned with rural development have estimated poverty's prevalence. This is probably for political reasons. We consider the rural population whose average daily calorie intake is estimated at 90 per cent or less of daily requirements to be poor. According to FAO and WHO recommended criteria, this group is at high risk of malnutrition.[16] Among rural households sampled in 1976, 28 per cent fell into the category of low-income groups, based on an average-sized rural household of 7.8 persons. Because of a wide variation in actual food intake within and between members of households based on age, sex, health and working conditions, this does not mean that all individuals in these groups were undernourished. Considering the small size of the subsample surveyed in 1976 and allowing for errors in underestimating the non-cash expenditure on food from self-produced items (cereals, dates, milk and fish), we roughly estimate the order of magnitude of the rural poor at 15–20 per cent in 1976. They have living characteristics consistent with those of poverty.[17]

Cuba

Prior to the 1959 land reform, 9 per cent of the landowners held 73 per cent of the arable land. American-owned sugar companies controlled 23 per cent of the total agricultural land. Sugar cane represented 53 per cent of the total crop area, with sugar and cattle combined amounting to 52 per cent of the total agricultural product. At the same time, food imports accounted for 23 per cent of total imports (Gayoso 1970: 19–22). In May 1959, with clear economic objectives of sharply reducing inequality, abolishing tenancy, diversifying and intensifying agricultural production, land reform was introduced. Productive assets and educational opportunities were also redistributed. To reduce dependency on sugar cane and to increase food production, the institutions of private farms, production cooperatives and state farms were chosen under a comprehensive planning system. This path was reinforced by according agriculture and expansion of irrigation priority. Accordingly, irrigated land was expanded from 13 per cent in 1959 to 32 per cent of total arable land in 1984. Fertilizer consumption rose sharply from 63 kg per hectare in 1960 to 179 kg in 1984.

Expropriated private land in excess of 67 hectares was redistributed to the peasants in units of 30 hectares (on average) free of charge as private and individual farms. Large sugar cane plantations, cattle ranches and unused land were converted into state farms (people's estates), the area of which amounted to 60 per cent of total agricultural land in 1963. This policy choice, and the ability to implement it, were studied by me in August 1959 when I examined the thinking behind it with Mr Antonio Jimenez, President of the Agrarian Reform Organization, and Dr Regino Boti, Minister of Economy.[18] Their view was that the reform aimed at liberating the tenants, sharecroppers and landless workers from exploitation by large absentee farmers and American monopolists in agriculture. Private property in land was granted as family farms (almost free), and state farms were considered as the engine of agricultural growth and the focus of socialist ideology. Production cooperatives (collectives) were considered unwarranted in the view of the two Cuban authorities because of the already large size of private farms (average 35 hectares) after the 1959 land reform. Instead, these private farms were to be controlled by the state through such means as the supply of inputs, pricing and marketing procedures by service cooperatives. The situation changed substantially following the second reform in 1963, which lowered the size of distributed units to 16 hectares on average and expropriated an additional 25 per cent of private farms of over 67 hectares each. The expropriated land was taken over by the state to expand the dominant state farms. Furthermore, those who owned private farms were encouraged to affiliate with state farms. Production cooperatives (collectives) grew from less than 1 per cent of total agricultural land in 1963 to 6 per cent in 1985–86, and state farm areas increased from 60 per cent in 1963 to 83 per cent in 1985–86.

Impact on agricultural growth and equitable rural development

Since 1959, these continuing institutional reforms have strengthened the command of the state in many aspects of managing the Cuban rural economy: organization of agricultural production, supply of the means of production, marketing of the products and processing of agricultural raw material – all governed by a comprehensive planning system. Production of livestock and crops grew at a fast rate despite a number of problems – the drought of 1961, the hurricane of 1963, the emigration of many technical staff to the USA, and a series of ideological uncertainties during the 1960s with regard to material incentives – all of which disrupted the output level of sugar cane until 1965.[19] The index of agricultural production (1974–76 = 100) grew at 3.9 per cent in the 1960s and 3.4 per cent in the 1970s. Notable was the fast growth in cereals by 12.3 per cent in 1970–80 (FAO 1985c). There was also a rise in value productivity per agricultural worker in the state farm sector at constant prices between 1971 and 1983 from 1,430 to 2,180 pesos according to a study by the International Labour Organization of the UN (ILO) (Ghai *et al.* 1988). This reported that between 1970 and 1983 total investment in agriculture increased by 130 per cent, fertilizer consumption by 34 per cent and irrigated area by 73 per cent (Ghai *et al.* 1988: Table 6.13). It stated that, while sugar retained its grip over the export sector, crop production for domestic consumption consistently increased its relative share.

With the second 1963 reform, resulting in a ratio between the maximum size of private landed property to distributed units of 2:1 (67 and 35 hectares on average), the provision of material incentives, food subsidies, rapid expansion of education free of charge and social benefits, greater equality in rural income distribution was realized. Measured by the Gini coefficient, inequality declined from its post-land reform low level of 0.35 in 1962 to 0.28 in 1973 and declined further to 0.21 in 1978 (Ghai *et al.* 1988: Table 6.7). With the massive redistribution of land and income, these characteristics of a highly egalitarian rural society suggest a minimal level of poverty, moving towards complete elimination.

From the findings of three recent studies, it appears that this remarkable contribution of Cuba's complete land reform has continued into the 1990s (Barraclough 1973, 1999; Deere *et al.* 1995). The rapid advances in the elimination of absolute poverty, egalitarian distribution of income and improvement in health, education and nutritional standards have taken place over the past four decades, despite the collapse in the early 1990s of the Soviet Union, on which the small economy of Cuba has depended, and in spite of the USA-imposed trade embargo, particularly since its 1992 Act. According to FAO statistics, the total food population index (1999–2001 = 100) increased from 105 in 2000 to 108 in 2003, and per capita food production also increased from 105 to 107 during the same period. Yet, producers' incentives have been increased by reorganizing the huge state farm sector in 1994–95, considering that production cooperatives have been linked to the industrial sector by the long-maintained socialist development planning (Deere *et al.* 1998).

Countries having quasi-complete land reform

Mexico

Land reform policy in Mexico has been a burning issue since the early twentieth century. It has been evoked by the semi-serf relations of forced labour within a socially semi-feudal agrarian system superimposed by the Spanish colonizers on the ethnic structure. However, the instituted scale of 1915 and 1971 land reforms and the resulting pattern of agricultural growth have fallen short of the expectations of the large indigenous population. The first task of land reform was to abolish compulsory unpaid labour and land grabbing and to restore and protect the *indios'* property rights in land.

The 1915 and 1971 land reforms

The most unique feature of Mexico's land reform is that it has been a continual process since the first law of 1915, fixing neither a date for completion nor a clearly fixed maximum limit on privately owned land. It was later, in the Land Reform Act of 1971, that the ceiling was defined as 100 hectares per person (with exceptions up to 300 hectares depending on the cropping system) of irrigated land, 200 hectares of rainfed land and up to 50,000 hectares of arid pasture land. The traditional system of communal ownership and use of land was institutionalized by *ejidos* and *comuneros* for the indigenous population.[20] For their protection, their communal property is inalienable; it cannot be sold, mortgaged or rented to others. On average, each household received between 2 and 5 hectares of irrigated land, and many did not pay for it. Thus, the foundation of inequality was laid down by the established ratio of maximum (the ceiling) to allocated irrigated land of around 40 or 50:1. According to the 1970 agricultural census, only 17 per cent of the *ejido* land was classified as cropped area (the remainder was undeveloped pasture and other types). Despite a lower quality of soil fertility, cropland in the reformed sector saw a 113 per cent increase in land productivity in 1950–70. Dovring (1970) found a wide production gap between reform and non-reform sectors over the period 1940–60 due partly to the differential increase in the proportion of cropped area in these two sectors.

This feature of dualism can be explained by the regional distribution of public investment in irrigation. The data from the agricultural censuses of 1950, 1960 and 1970 show that irrigation investment has been allocated disproportionately by the government with much larger shares in the states of Sonora, Sinaloa and others in the North Pacific region where large commercial farmers dominate in alliance with MNCs. We should note that the Mexican Constitution forbids foreigners to own agricultural land. The data imply greater investment in high-value cash crops, such as soya, oil seeds, fruits and sugar cane, which are grown predominantly in the northern states. In the southern regions (Gulf south, Peninsula and the Central Plateau), where *ejidos* and traditional Indian villages (*comuneros*) are dominant, nearly 75 per cent of the cultivated land grows mostly beans and corn, the two Mexican staples.

Sustained food production

Contrary to the popular belief that reduced food production necessarily follows land redistribution, total and per capita food production grew rapidly in the 1950s and 1960s, despite the fact that the reform areas represented about 40 per cent of total cultivated land and the number of *ejidotarios* almost half of the total agricultural households. In Table 5.10, we present the FAO index of food production (1952–56 = 100), which clearly shows this sustained growth.

Even in the earlier period of land reform (1930–40), Dovring's study concludes that 'there is no evidence to show that the early land reform measures had any negative effect on production' (Dovring 1970: 35). With the institution of the *ejido*, land reform's contribution to the indigenous population has been in its provision of an alternative to continued poverty or outmigration of the rural poor. Through intensified family labour and the *ejidos*' response to technological change via their application of high-yield varieties of wheat and corn, the indigenous households have gained command over their food needs and, through granting inalienable property rights, land reform has ensured permanent security of tenure for community members against outsiders, including moneylenders.

Table 5.10 Post-reform index of food production in Mexico, 1956–68

	Total food production	Per capita production
1956	116	109
1957	132	120
1958	142	125
1959	140	119
1960	145	119
1961	154	123
1962	158	122
1963	165	123
1964	175	126
1965	182	125
1966	186	125
1967	199	129
1968	206	129

Source: SOFA (1970: Annex Table 6c).

Tunisia

This is the second case study of the quasi-complete land reform group of countries. It is selected according to the classification suggested earlier and the data presented in Table 5.1. We begin with an understanding of its brief historical context, which suggests that the question of holding agricultural land and its political and social implications has been central to the historical experience of Tunisia since the Arabs ousted the Romans in the seventh century. In consequence, the principles of inheritance, land tax (*ushr* and *Kharaj*) and holding landed property under the Islamic *waqf* (*habous*) were introduced. Subsequently, during the long Ottoman rule (1569–1881), land taxation was excessively increased around the middle of the nineteenth century, which triggered rural unrest and the *fellaheen's* revolt in 1864 (Perkins 1984: xi, 8). The economic situation was worsened by devastated cereal crops and the periodic outbreak of cholera and famine in rural areas in the 1860s. This human disaster forced many *fellaheen* to flee to towns, and pressed upon the rulers to borrow heavily from French financing institutions (Issawi 1982: 97–100). The coalition between lending institutions in France and other European creditors influenced the French government to occupy Tunisia, which was declared a Protectorate in 1881.

Duality, inequality and poverty prior to land reform

By 1956, when Tunisia gained independence from French rule, some 800,000 hectares of fertile land (20 per cent of total arable land) was held by 6,500 Europeans, mostly French. Out of this total area, 600,000 hectares was owned, and the rest was held under rental arrangements. These actions formed a modern sector, in which the average size of farms ranged from 100 to 200 hectares held by a single French settler. They were predominantly located in high-quality rainfed lands where the mechanized wheat farms, modern irrigation methods and high-value vegetable crops and fruits were introduced.

Reliable data on living conditions in rural areas at the time of independence are scarce. The census of 1956 tells us that the population was overwhelmingly rural (70 per cent), and 57 per cent of the total workforce depended on agriculture for employment. Chebil (1967) reported that, although there were a few modern Tunisian farms whose owners were rich, most of the Muslim Tunisian workforce in agriculture (0.7 million) were poor. The rural poor comprised small landholders in rainfed areas, sharecropping peasants, landless workers and nomadic pastoralists. Nearly half the workforce in agriculture was landless workers.

About 1 million hectares, not alienable in the land market, was operated under the Islamic land tenure arrangements of *waqf* (*habous*), poorly managed and mostly tilled by insecure tenants and landless squatters. Approximately 2.5 million hectares of marginal lands in the central and southern regions was held customarily by semi-nomadic tribes under communal ownership. Chebil (1967) also reported that poverty prevailed in the 1950s among the *fellaheen* in the *habous* lands and in areas of lands rented out by Muslim Tunisians. These agrarian conditions were manifested in low annual income per working person in agriculture, estimated in 1966 by the World Bank at 84 Tunisians dinars (about US$210 in 1966), or less than one-fifth of that of the non-agricultural labour force. Several estimates were made of the prevalence of rural poverty in 1966, ranging between 49 per cent and 60 per cent (Radwan *et al.* 1991: Table 4.10).

Reforming the land tenure system, 1956–69

The reform began with the abolition of the archaic institution of public *waqf* or *habous* by laws of 1956 and 1957. The state acquired their property and abolished the Ministry of *Waqf.* Private *habous* lands were divided into individual owner-ship units among the legitimate claimants. This reform affected 29 per cent of total arable land or 4.2 million hectares, excluding the areas of forest and natural pasture. This extensive area was gradually transformed into productive farming enterprises through irrigation investment and improvement of cereal and livestock production. Next came the regulation of the communally held tribal land rights by laws of 1957 and 1959. Most of these hitherto underutilized lands were changed to individual ownership in units of 10–20 hectares each, mostly with olive planta-tions, and the rest remained under communal ownership. Government-patronized production and service cooperatives were organized by a law of May 1963, pro-viding beneficiaries with technical assistance, construction of tube-wells and the basic means of production to diversify and intensify land use according to rainfall. Certificates of possession were granted by the state to individual landholders, giv-ing them the right to borrow from the state-owned agricultural credit bank.

Limiting the maximum size of landownership at 50 hectares was confined to the newly irrigated areas of the Mdjerdah Valley scheme by the law of June 1958. The area exceeding that limit was expropriated against payment of compensation in the form of guaranteed Treasury bonds. It was redistributed in ownership units of 5–10 hectares to small tenants with experience in irrigated farming. The crite-rion for unit size was to provide a minimum income of 250 dinars per family per year (US$600 in 1965) from growing vegetables, fruits, wheat and from dairying (Parsons 1965). The new owners were required to pay for the land over a period of 20 years and to join production cooperatives, which became the sole source of credit and the agent for marketing the products. One notable characteristic of this policy is the requirement that landowners share in the costs of publicly funded irrigation works, according to land productivity and its value. The charge ranged from 25 per cent to 60 per cent of value added (Chebil 1967: 196).

An important instituted agrarian change was the scheme for the consolida-tion of small scattered parcels of land. This scheme and the above-mentioned three land reform acts have laid the foundation for technological change and rural development. Apart from the mandatory membership of cooperatives, the govern-ment provided the rural population with primary education for boys and girls, health and child care services, housing, adult literacy and skills formation for the youth. After the consolidation of fragmented holdings came the recovery of the commercialized foreign-owned farms, 8 years after independence, by force

of the Agricultural Land Property Law of 1964; nearly 16 per cent of total cultivable lands or 700,000 hectares of the best land in the country was taken over by
the government. The acquired land was partly purchased at the market price and
partly nationalized. A small portion of it was allotted to landless farmers, and the
rest was retained by the government and managed as state farms. Throughout the
countryside, farmers holding individual ownerships (*melk*) and leased lands were
grouped into collective production cooperatives, which were virtually managed
by a hierarchy of bureaucracy. Experience has shown that the hurriedly implemented mandatory cooperativization did not work, particularly where individual
land titles were granted. They were short lived, and in 1969 were dissolved after
widespread and growing discontent among farmers with the unsatisfactory results
of this imposed system. Since 1969, voluntary membership has replaced compulsory adherence, and individual management responsibility of private farms has
prevailed.

Distributive consequences

The above account of the institutional transformation during the period 1956–70
illustrates the effects on production and income distribution in rural Tunisia. From
my discussions during the 1976 field study with both cooperative leaders and
farmers about the poor production performance in the 1960s, two explanations
were given. One was the droughts of 1961, 1962 and 1967, and the other was the
government experimentation with collective cooperatives, which were virtually
imposed on farmers. The agricultural stagnation during the 1960s has ended, and
agricultural production improved in the 1970s according to my field observations
in 1985 after the rapid irrigation expansion that took place between 1970 and
1985. This technological advance has increased the irrigated land area by 160 per
cent, and the agricultural GDP rate of growth has doubled to 4.2 per cent compared with 1960–70. New owners' security of tenure has induced their favourable
production response to this technology and to the widely introduced high-yielding
varieties of wheat.

The 1980/81 survey of landholdings conducted by the Ministry of Agriculture
shows that 85 per cent of the total farming population had access to land (owners, tenants, sharecroppers and agro-pastoralists). Nearly one-fifth of agricultural
families were left out by the several agrarian reforms as landless wage workers.
Since then, some have migrated to work in oil-rich Libya, and some have been
absorbed into the government-promoted non-agricultural activities within rural
areas as well as the expanding urban centres of the economy. The results of the
survey also show that, out of the total arable land, the major part (87.6 per cent) is
privately owned (*melk*) in units of 13 hectares, on average.

The National Institute of Statistics (INS) analysed the results of the 1981 survey and reported that small farmers in the size category of less than 10 hectares
represented 63 per cent of the total number of farmers. The INS (1982: 58) also
reported that the landless workers representing 14 per cent of total farming house-

holds in 1981 had the highest prevalence of poverty (37 per cent) and illiteracy (77 per cent). Two different estimates were made for poverty levels: the ILO team of experts at 42 per cent for 1980 and 31 per cent in 1985, whereas those by the government INS were 14 and 7 per cent respectively (see the details of estimates and different methodology used in El-Ghonemy 1993: 136–7, Table 7.4).

6 Case studies II

Partial land reform

My first point is to remind the reader of the classification of countries' experiences in state-mandated redistributive land reform into complete, quasi-complete and partial land reform, as defined in the preceding chapter. The first two categories have already been presented, and the third category is examined in this chapter. In alphabetical order, the countries studied are Egypt, India, Iran, Morocco, the Philippines and Syria. Had space permitted, Chile, Pakistan and Sri Lanka would have been included. But before presenting each case study, some general remarks are made with regard to the features of partial land reform (PLR) that are relevant to the subject of this book.

Some common features

The redistributive scale is the primary criterion used in our classification, and ranges between 45 per cent and 4 per cent for land area out of the total agricultural land area and between 45 per cent and nearly 2 per cent for the beneficiary households out of the total agricultural households. It seems that the leadership in PLR countries preferred to limit the land reform scope through granting high ceilings, in the hope of minimizing damage to landlords and maintaining their political support. In the marketing of crops and related credit supply, PLR's control of mercantile interests is also relaxed. In broad terms, economically powerful members of rural communities continue to dominate local institutions and political organizations. This dominant role is likely to be manifested in the working of the land, labour and capital markets.

Furthermore, PLR countries share a common distinct duality in the agrarian economy. Whereas complete land reform brings about an egalitarian agrarian structure, partial reform divides the agrarian structure into two major subsectors: the created reform sector and the non-reform sector. The production and marketing implications of such duality start from the differential productive quality of land in the reform sector compared with the non-reform sector. Given the time and freedom to choose the land area below the fixed ceiling and capital equipment to retain, landlords often hold on to the best, while the expropriated portion is usually less productive. Furthermore, affected landlords keep substantial capital in proportion to the area left after expropriation. This has implications for the organization of production and related technological change; the employment effect is likely to be small as landless workers are often excluded in preference to tenants actually cultivating the land, and new owners tend to rely on family labour for cultivating their own landholdings.

The power structure is likely to be determined by the residue, i.e. the extent of both the land remaining as private property after curtailing the large estates and the economic activities of the affected landowners, traders and moneylenders, as well as their influence exerted on the decision-making structure at local and national levels. In the next chapter, we shall learn how, in the post-economic reforms of the 1980s and 1990s, these features of the power structure have weakened the PLR beneficiaries' gains, particularly with regard to affected landowners' lobbying strength for the removal of legal provisions, protecting small peasants and hired agricultural workers from the undesirable effects of free market forces (e.g. eviction of tenants, resumption of their pre-PLR monopoly powers in the labour, credit and land markets).

Egypt

I have examined elsewhere (El-Ghonemy 1953) the characteristics of Egypt's rural economy during the decades immediately preceding the land reform of September 1952, which are briefly explained as follows:

1 There was a heavy concentration of landholdings and extensive absentee landownership. Thirty-five per cent of agricultural land was possessed by 94 per cent of the total number of landowners, whereas 0.1 per cent of total owners controlled, among themselves, 20 per cent of the Nile valley's fertile land. A process of polarization rapidly developed, whereby the percentage of the mini-owners of less than 2 hectares (5 feddans) increased, while the average size of their ownership decreased, from 0.42 to 0.3 hectares, between 1916 and 1950. Accordingly, income distribution became highly skewed. Average annual net farm income of the owners of less than 1 feddan (0.4 hectare) was only 0.05 per cent of that of those rich landowners in the size group of 200 feddans (83.3 hectare) and over in 1949/50. Measured in terms of the Gini index, this high inequality of income distribution was 0.858 (El-Ghonemy 1953: Table 35).

2 About 60 per cent of the total area of cultivated land was operated under insecure tenancy arrangements. The percentage was even higher, at 86 per cent, in some provinces where large estates and foreign-owned plantations of sugar cane and cotton existed. Landlords and foreign enterprises exercised monopoly power in their locations and received high monopoly profit. They established institutional barriers for the direct lease of land, and also had the right to evict tenants without compensation payments. By leasing out land through auction, they forced small tenants to rent from a hierarchy of intermediaries, resulting in the payment of exorbitant rents. Consequently,

rental values rose fourfold in real terms between 1931 and 1950. Combined with capital rationing in the institutionally provided credit union (which was primarily controlled by two foreign banks), these high rents resulted in both the chronic indebtedness of the tenants and very low incentives for investment to improve land productivity.

3 Despite substantial improvement in the irrigation system and cotton yield, the rural economy experienced a steady decline in productivity per agricultural worker due to the fast-growing agricultural population and the highly inelastic supply of cultivated land. Land concentration, combined with demographic pressure, led to a rise in landlessness (agricultural households who neither owned nor rented land), which I estimated in 1950 at 45 per cent. A Malthusian situation existed: food productivity per capita of total population in 1949–52 was 70 per cent of its 1900–04 level; and population pressure on the land increased, resulting in a fall in the ratio of cultivated land per head of agricultural population from 0.78 acres in 1900 to 0.44 in 1952.

4 In the face of low wages, high rental values, rationing of the credit market and inflationary land prices, the mass of hired farmworkers and small tenants did not have the earning capacity to purchase a piece of land. Nor were they likely, for social reasons, to obtain land through institutional means such as marriage, inheritance or as gifts. The sale of reclaimed state-owned land in public auction was largely captured by the big landowners and foreign land companies (90.7 per cent of the total reclaimed land area during 1935–50).

5 Social unrest and discontent prevailed in the countryside among the indebted tenants and the unorganized poor farmworkers. Several proposals to improve the land tenure system (particularly after the Second World War) were silenced by the strong coalition of the dominant influential landlords, members of parliament and the former royal family, who were themselves the largest landowners in the country.

Thus, the pre-1952 land-based power structure obstructed rural development and stratified the rural society into an upper-class minority of rich landlords and cotton merchants and a mass of very low-income and poor *fellaheen*. In between lay a wide social gulf. Hence, the existing pattern of land distribution, other productive assets and educational opportunities, and income led to widespread poverty, the magnitude of which I estimate at nearly 56.1 per cent of the total rural population in 1950 based on annual income levels. My estimate is a result of several studies carried out in villages during the period 1947–50.[1]

The contents of 1952 and 1961 partial land reform

Against this background, the September 1952 land reform was instituted by the revolution of July 1952 led by Gamal Abdul Nasser. This radical institutional change was a response to simultaneous calls for reform by a group of Egyptian intellectuals and an outbreak of rural unrest in several localities. When the land reform was proclaimed, its overriding objectives were social and political: to build

Egyptian society on a new basis by providing free life and dignity to each peasant, by narrowing the wide gap between classes and by removing an important cause of social and political instability (translated explanatory note to Land Reform Act No. 178, 9 September 1952, in Arabic). Economic goals were broadly expressed in terms of 'to pave the road for rapid economic development' and 'to check the rush of those rich possessing savings to purchase land and to direct every new investment into reclamation of new lands, industrial and commercial activities'.

But these bold objectives and the pre-1952 aspirations of peasants and social reformers among the intelligentsia were only partially realized by the series of land reforms issued between 1952 and 1964 (Table 6.1). By 1980, after the completion of their implementation, 847,887 feddans (356,112 hectares) or 13.9 per cent of the total cultivated land area was redistributed to 353,286 peasant families representing 9.6 per cent of the total number of agricultural households. This redistributed area represented 87 per cent of the total land acquired by the state through confiscation, expropriation against compensation payments and purchases from foreign-owned land companies. The balance, mostly orchards and less productive land, was retained by the government as the property of the Agrarian Reform Authority according to Law No. 52 of 1966. In addition, 214,166 feddans (89,949 hectares) of state-owned reclaimed land, using the Aswan High Dam water, was distributed during the 1960s (Egypt's Central Statistical Organization *Statistical Yearbook*, June 1986: Table 2-22).

Following completion of land reform implementation, between 30 and 38 per cent of the total number of agricultural households remained wage-dependent landless workers, and they suffered the consequences of being excluded from the reforms. The prescribed minimum daily wage rates in the 1952 law remained ineffective until the labour market allowed their upward trend in the late 1960s and 1970s. Furthermore, cultivated land under tenancy arrangements was marginally

Table 6.1 Areas acquired for redistribution by Egyptian land reform laws, 1952–69

Land reform laws	Area in feddans
Initial Land Reform Law No. 178 of 1952, and Law No. 598 of 1953 for confiscation of ex-Royal family estates	450,305
Law No. 152 of 1957 and law No. 44 of 1962 for the transfer to land reform of *waqf* lands entrusted to charitable and public purposes	148,787
Second Land Reform Law No. 127 of 1961 for reducing maximum landownership of individual household to 100 feddans	214,132
Purchases of lands sequestrated in 1956 including those of Kom Ombo Land Company	28,807
Law No. 15 of 1963 for the acquisition of foreigners' landownerships	61,910
Law No. 150 of 1964 for the confiscation of land owned by Egyptians put under sequestration (Hirasah)	43,516
1969 law amending some of the above laws (my estimate)	32,000
Total	979,457

Source: Compiled by the author from the records of the Agrarian Reform Authority, Ministry of Agriculture, Cairo.

reduced from 60 per cent in 1951 to 50 per cent in 1965 because tenancy was regulated by fixing a ceiling on rental values at seven times the land taxes that were assessed in 1949 and at 50 per cent of harvest in the case of sharecropping. The 1952 law provided the tenants with security of tenure for a renewable period of 3 years under written contract registered at the village cooperative. Auctioneering land for lease was prohibited, as was subleasing. The maximum area rented by any cultivator was limited to 50 feddans (20 hectares).

The initial land reform of 1952 fixed a size ceiling of 300 feddans or 126 hectares on land privately owned by one household, while that of 1961 lowered this ceiling to 100 feddans or acres (42 hectares). In 1969, a ceiling of 50 feddans was fixed for one person's ownership, but its effect was minimal. Land acquired by the state was to be redistributed to tenants (already cultivating the land) in small units of individual ownership of 2–3 acres (about 1 hectare on average). Table 6.2 shows the calculated percentage size distribution of landownership before and after land reforms. We need to keep in mind that, throughout the reform process, the institution of private property rights in land has been maintained in practice and established in the New Constitution of 1956 and the Egyptian Charter of 1962 (Mithaq). With political will and high implementation capability, the programme of land redistribution progressed on schedule.[2] Over a period of 40 years, state institutions were to compensate (in government bonds) affected landowners whose land was expropriated against instalment payments by the new owners. In turn, new owners were charged 15 per cent of the land price for administrative costs (later reduced to 10 per cent). In 1964, land recipients were exempted from the payment of land tax.

Egypt's partial land reform is manifested in the ratio of the fixed ceiling to the average size of the units redistributed to new owners. At a ratio of 126:1 established by the 1952 land reform, the ratio dropped to 40:1 by the 1961 re-

Table 6.2 Changes in the size distribution of landownership in Egypt, 1951–84

Size of ownerships (feddans)	1951		1965		1984	
	0%	A%	0%	A%	0%	A%
< 5	94.3	35.4	95.0	57.1	95.2	53.0
5–10	2.8	8.8	2.5	0.5	2.5	10.4
10–20	1.7	10.7	1.3	8.2	1.3	10.9
20–50	0.8	10.9	0.9	12.6	0.7	11.9
50–100	0.2	7.2	0.2	6.1	0.2	7.5
100 and over	0.2	27.0	0.1	6.5	0.1	6.3
Gini coefficient of landownership	0.611		0.383		0.432	
Gini coefficient of landholdings	0.715 (1950)	0.456 (1975)				

Notes
0% is the number of ownerships as a percentage.
A% is the area of ownership units as a percentage.
One feddan equals 1.04 acres or 0.42 hectares.

form. Another measure of the partial scope is the changing share of size classes of landownerships and the corresponding share in land shown in Table 6.2. The size class of most of the individual household ownerships exceeding 100 feddans (40 hectares) has been almost eliminated. Also, the share in total land of 95 per cent of the small owners of less than 5 feddans (2.1 hectares) increased from 35.4 per cent before the reforms to 57.1 per cent in 1965. But the shares of the size category of 10–50 feddans (4–20 hectares) in total landownerships remained virtually stable, as did the upper class of 50–100 feddans (20–40 hectares), whose percentage remained unchanged at 0.2 while the corresponding share in total land fell by only 0.9 per cent.[3] Thus, after a series of land reforms, the degree of land concentration was improved substantially, but there is still high inequality and the number and area of the dwarf holdings of less than 1 acre doubled in one decade, between 1950 and 1961, suggesting a rapid process of fragmentation. In the following discussion, the changes cover the period 1952 until the late 1980s, leaving the post-economic liberalization changes of early 1990 to Chapter 7.

In the almost rainless Egyptian agriculture, the government has always been in full control of the management of the Nile water supply for irrigation, a policy reinforced with the huge public investment in the construction of the Aswan High Dam in the late 1950s and early 1960s. The state control of planning crop rotation has been further extended from the reform to the non-reform sector. In addition, regulating prices and the marketing of main crops have also been controlled throughout agriculture until around 1995, following the implementation of structural adjustment and price liberalization (see the early 1990s changes in El-Ghonemy 2003: Ch. 6).

Agricultural output, employment and productivity

The organization of production within the crop rotation presents a special feature of Egyptian land reform. Redistributed units are divided into two or three separate pieces to match the planned crop rotation in each cooperative's area. The amalgamated pieces constitute a large plot (hode) of 30–50 feddans (12–20 hectares) planted by a single crop (crops are then rotated on a 2- or 3-year cycle). Each owner has a distinct area registered in his or her name, cultivated individually, and the produce of the land is so identified after harvesting. This innovative arrangement combines the advantages of both private property incentives and large-scale production. Small units are consolidated to reap the benefits of the agronomic–technological advances and can be ploughed, irrigated and sprayed against insects at the same time. A significant impact of land reform on the large non-reform sector was the forceful and rapid extension of this system in 1964–66 from the land reform subsector to the rest of the agricultural sector. In 1956, Professor Mario Bandini of Italy (on whose advice this system was initiated in Egyptian land reform) expressed his view to the present author that it represented a pragmatic solution to the distributed small units along with the availability of a large number of professionally trained university graduates in agriculture. There is no scarcity of studies about the impact of land reform.[4]

Because the institutional organization described above was confined to land reform areas until 1964 and only then extended to the rest of the farming areas, comparable data on average productivity per unit of land during 1952–64 are more relevant. I have compiled case studies carried out during this period by the Land Reform Authority (*Hai'at Islah Zira'i*) in three areas (El-Ghonemy 1968a), and they are reproduced in Table 6.3. The data indicate that the yields of cotton, sugar cane and maize have clearly risen at faster rates in the land reform sector than in the national averages, while rice yield was uniformly high. At the national level, total food production grew in 1952–64 at the annual rate of 3.5 per cent and per capita production at 1.1 per cent, and the annual growth of total agricultural production increased from 2.8 per cent in 1952–59 to 3 per cent in 1961–70. Hansen and Marzouk (1965: 78) estimated that, during the initial phase of land reform (1952–60), net value added productivity per feddan increased in real terms by 30 per cent and per unit of agricultural labour force by 21 per cent (consisting of 11 per cent value added productivity and 10 per cent accrued from the transfer of value added from absentee owners excluded from the labour force to the active owners and tenants actually cultivating the land).

The same variation between land productivity in the reformed and non-reformed subsectors was confirmed in my sample survey conducted in 1973 in three reformed areas (Inshas, Itay Al-barood and Gabaris) located in two provinces in the Nile Delta.[5] The sample covered 611 households and compared their 1973 social conditions with the initial data collected in 1953/4 at the time of land reform implementation; the comparison between the two years in Gabaris is presented in Table 6.4. However, the 1973 data for the three reformed areas with regard to the main crops' average yields compared with average national yield are shown in Table 6.5. They indicate higher cotton yields than the national average while those of other crops are almost equal.

Table 6.4 on the beneficiary households' main changes over a period of 20 years (1953–73) shows a rise in the household size of new owners from six to eight persons, while their land remained constant. However, thanks to the rise in their livestock assets and in areas cultivated by vegetables, fruits and clover (*barseem*), the gross per capita value of output increased. We do not know how the household gross income was estimated in the initial year 1953/4, when the beneficiaries were allotted ownerships. The increase in livestock assets and their products represents a significant change in the new owners' income, reaching 44 per cent in the Inshas area. And, on average, family labour accounts for nearly half the total labour inputs in cotton and 70 per cent in rice production, with hiring-in at its highest during the peak season of cotton picking and cutting and rice planting and transplanting.

In sum, the evidence presented suggests that: (i) land reform beneficiaries with small units of land (2.4–2.8 acres) all hire in labour during the peak seasons; (ii) the share of crop value in gross household income has diminished while that of livestock ownership and non-agricultural activities has increased significantly; and (iii) sampled areas show that not every land reform area has higher yields than the rest of the agricultural sector.

Table 6.3 Average yields per feddan in three land reform districts and national averages, Egypt 1954–64

District	Main crop	Unit	Average yield per feddan 1950/52		Average yield after the implementation of 1952 land reform											
			Local	N	1953/54		1955/56		1957/58		1959/60		1962/63		1963/64	
					LR	N	LR	N	LR	N	LR	N	LR	N	LR	N
El Manshia (cotton area)	Cotton[a]	Kintar[b]	3.5	4.2	4.1	3.9	3.6	3.5	4.7	4.7	6.0	4.2	6.8	5.4	7.1	8.0
	Wheat	Ardab[c]	4.1	5.1	4.3	6.4	5.7	6.6	4.8	6.6	5.7	6.9	8.0	7.3	6.3	7.7
	Maize	Ardab	4.7	6.3	4.7	6.5	6.6	6.4	6.9	6.4	8.9	6.3	8.9	6.9	10.6	7.8
	Cotton	Kintar	3.5	4.2	5.0	3.9	4.0	3.5	5.4	4.7	3.3	4.2	5.5	5.4	5.4	6.0
Dernera (rice area)	Rice	Dariba[d]	1.7	1.4	2.7	1.9	3.0	2.3	10	2.1	2.7	2.2	3.2	2.6	3.2	2.3
	Maize	Ardab	4.5	6.3	6.5	6.5	7.0	6.4	8.5	6.4	10.5	6.3	11.0	6.9	–	–
El Mata'ana (sugar cane area)	Sugar cane	Ton	37.0	39.0	45.5	41.7	52.0	41.4	57.5	41.5	51.3	45.5	51.2	44.1	52.0	36.5

Source: Based on data compiled from studies carried out by the Statistics and Research Office of the Land Reform Authority, Dokki, Giza. National averages are from *Bulletin of Agricultural Economy*, Ministry of Agriculture, Dokki, Giza (several issues).

Notes
a Long staple cotton.
b Kintar = 157.7 kg.
c Ardab = 150 kg.
d Dariba = 954 kg. Ton = 1,000 kg.
LR, average yield in land reform areas.
N, national average yield.
Data on national average yield for cotton show the yield per same variety grown as in land reform areas.

Table 6.4 Changes in household asset ownership and per capita income in a land reform
area, Gabaris, Egypt, 1953–73

	1953/4	1973	Change (%)
Number of new owners' households sampled	110	110	–
Number of persons	690	981	+42
Average size of household	6.2	8.9	+44
Land per person (feddan)	0.45	0.31	–31
Livestock (head per household)	1.0	2.6	+160
Estimated average gross income per household (pounds, current prices)	73	515[a]	+605
Per capita real gross income (adjusted by consumer price index for food 1953 = 100)	12.5	22.7[b]	+816.6

Source: Calculated from a sample of 110 new owners (5 per cent at random) conducted by the author
in 1973 in collaboration with the Egyptian Land Reform Authority.

Notes
a The share of value added by livestock products is 41 per cent.
b The consumer price index was 244 in 1973.

Table 6.5 Average yield of main crops in three reformed areas of Egypt compared with
the national average

	Inshas	Itay Al-barood	Gabaris	National average
Cotton (kintar)	5.5	6.0	6.8	4.6
Wheat (ardab)	11.0	9.4	9.9	9.8
Maize (ardab)	10.0	10.8	10.9	10.0
Rice (ton)	2.4	2.2	2.5	2.3

Distribution of income effects

We have seen from the data given in Table 6.2 a substantial reduction in land concentration in terms of the Gini index between 1951 and 1965, primarily resulting
from the set of land reform laws. Once enforced, instituted rent control had an
even greater impact as it benefited all tenant cultivators, who, in the 1961 agricultural census, were estimated at nearly 1 million landholders including sharecroppers. Costly and substantial food price subsidies providing an annual average gain
of E£25 on average per person did, however, reinforce the effects on income.[6] The
close association between redistribution of land and greater equality in the distribution of income and consumption is manifested by the sharp decline in the Gini
index of the latter from 0.858 in 1949/50 to 0.370 in 1958/9, and inequality fell
further to 0.270 in 1964/5. [The first was my own estimate and the latter is from
Adams (1985).] But the results of the official family income/expenditure survey
carried out in 1974/5, the ILO Rural Poverty Survey in 1977 and the IFPRI survey
in 1982 show a post-1965 rise in inequality. These different estimates are best
shown together in Table 6.6.

Table 6.6 Estimates of inequality (Gini coefficient) in income/expenditure distribution in rural areas of Egypt, 1940–82

1949/50	1958/9	1964/5	1974/5	1977	1981/2
0.858	0.370	0.270	0.348	0.393	0.337

Some explanations of the post-1965 rising inequality can be offered in conjunction with the changes in the rural labour market and the composition of asset ownership since mid-1970. Until Nasser's death in 1970, the land reform policy choice was accompanied by tight governmental control of the land and credit markets, as well as agricultural prices, food subsidies and the central planning of the entire economy. Around 1975, President Sadat's *Infitah* (opening or liberalization of economic activities) policy relaxed this control and, unintentionally, brought into force a chain of 'disequalizing' factors. They include the change in the profile of the rural labour market, which arose from the pro-emigration policy and dramatic changes in the international labour market within the Middle East. These changes induced landless workers and many skilled members of rural households to seek higher earnings in the neighbouring Arab countries, where the demand for labour was rising rapidly following the oil boom in 1973/4.[7] As observed by the present author in his own village, some of the sizeable remittances (reaching a total of US$3.9 billion in 1984 at the national level) were channelled into the land market, thus inflating the prices of the highly inelastic supply of land. Consequently, agricultural land prices rocketed, reaching about E£5,000 per feddan or US$6,000 per acre in 1984 (in contrast to E£800 for 1 feddan in 1975) according to the official exchange rate.

In addition, diminishing equality between 1965 and 1982 was also due to the adverse effect of land transactions among landholding groups, whereby owners in the size group of 5–20 acres (2–8 hectares) purchased land from those owners of 5 acres and less during the period 1964–84, as can easily be noted in the columns for the years 1965 and 1984 in Table 6.2. The study of Commander (1986: Table 8.24) showed that the landholders in the size group of 5–10 feddans were the highest gainers in the shares of gross income between 1977 and 1984. The differential is attributed to investment in rural enterprises such as raising poultry, animal husbandry for meat production and growing higher value crops such as vegetables and fruit. These enterprises were not subject to the government-controlled pricing system.

The alleviation of poverty can be judged from the results of the officially conducted Family Budget Surveys since 1958/9, the ILO Survey of 1977 and the IFPRI Budget Survey of 1982. They all point in the positive direction of a substantial improvement in nutritional level and a reduction in the prevalence of poverty. The results of the *Fifth World Food Survey* (FAO 1985a) confirm this positive change. The average daily calorie supply per person grew by 2.2 per cent per annum from 2,561 calories in 1969–71 to 3,178 in 1979–81. These improvements are reflected in the 50 per cent increase in life expectancy between 1950 and 1980. The estimates of the rural population falling below the poverty line corresponding to the successive years 1949/50 to 1982 show a proportionate decline from 56.1 per cent to 17.8 per cent during this period, i.e. 30 years after the initiation of land reform. The numbers of the poor have also diminished from 7.7 million to 4.2 million despite the rise in the average annual rate of population growth from 2.2 per cent to 2.8 per cent during this period. Although these are estimates based on different poverty lines and calculated average rural household size, they are – with these cautious comments – presented together in Table 6.7.[8]

With the exception of 1974/5, the estimates show a downward trend in poverty prevalence between 1950 and 1982. Sen's index is more sensitive to changes because it combines measurements of the percentage of people falling below poverty lines, the gap between their average income/expenditure and the poverty line, and the Gini coefficient of the distribution of income (or expenditure) among the poor. The last two measures may explain the rise in the index in 1964/5 and its sharp fall in 1982 when all three components seem to have improved. In fact, Adams (1985: Table 3) indicates a narrowing of the poverty gap in constant prices from E£32.5 in 1963 to E£23.4 in 1982.

This fall in rural poverty presents a paradox in the face of: (i) the limited scope of land reform in terms of the redistribution scale and (ii) the slow growth of agricultural output at the average annual rate of 2.7 per cent. Why, then, has the prevalence of poverty steadily declined? Apart from the initial but dramatic decline in the inequality of land distribution between 1952 and 1965 (Table 6.2), there is a complex of dynamic factors lying outside the domain of land reform (outlined above). They are centred around:

1   The externally induced change in the profile of the rural labour market and the consequential rises in real wages in agriculture and in the income share from non-land asset ownership (particularly livestock and farm machinery for hiring out).

Table 6.7 Prevalence of poverty in rural Egypt, 1949/50–1982

	1949/50	1958/9	1964/5	1974/5	1982
Rural population ('000s)	13,710	15,968	17,754	20,500	23,760
Percentage of rural poor	56.1	27.4	23.8	28.0	17.8
Number of rural poor ('000s)	7,691	4,468	4,225	5,740	4,229
Sen's index of poverty	–	0.079	0.178	–	0.061

Sources: 1949/50, El-Ghonemy, see Note 1 of this chapter on the database; 1958/9, 1964/5 and 1982, Adams (1985: 705–23), 1974/75, Radwan and Lee (1979: Table 5.3).

Note
For the meaning of Sen's index, see Sen (1981: Chapter 3 and Appendix C). The index for these years was calculated by Adams (1985). We should recall that land reform began in 1952.

2 Post-1973 food subsidies are substantial in their levels and in their coverage of consumer goods, which provided the poor with access to basic needs of cheap food, fuel and clothing.

3 Free education, leading to expanded access to employment opportunities outside agriculture.

India

Our review of state-administered land reform is incomplete without a brief reference to the rich experience of India in its partial redistributive policy as an integral part of tackling the problems of rural development. Here, national policy instruments have addressed most aspects of persistent poverty within a democratic process of policy formulation.[9] Throughout, the fundamental principles laid down in 1947 by the Congress Party Committee on Agrarian Reform have been maintained. Private property rights in land and other means of production have been adopted within a socially complex class system and in different states, some of which have populations exceeding 80 million.

Interstate variations in land reform scope

Constitutionally, each state makes independent decisions regarding the design of land reform policy and its implementation. Virtually all states have enacted numerous laws between 1952 and 1974, fixing ceilings on private landownership, abolishing intermediaries in tenancy arrangements, controlling rental values and fixing minimum wage rates in agriculture. Nevertheless, all these pieces of regulatory legislation are designed with such deliberate exemptions and legal loopholes that they could not meet the rising expectations of the mass of rural poor.[10] The end-result is that only 3 per cent of privately owned land in the whole of India has been redistributed.

Why has the fervour in advancing the cause of land reform since the early 1950s floundered in India? Clues can be found from a number of sources, including official statements and academic studies examined by the present author during his visits to India in the 1960s and 1970s, My interviews with senior officials in the Planning Commission and the Revenue Department of West Bengal and scholars at academic institutions have revealed the wide gap between idealism, on the one hand, and constraining inadequacies in implementation capabilities, on the other.[11] Bardhan succinctly summarized the constraints:

> These [land reform and tenancy control] laws were executed by a local bureaucracy largely indifferent, occasionally corrupt and biased in favour of the rural oligarchy. [. . .] Quite frequently, protective tenancy legislation may have worsened the conditions of tenants; it has led to the resumption of land by the landlords and eviction of tenants under the guise of 'voluntary' surrender of land.
>
> Bardhan (1974: 256)

Whatever the reasons might be, and ideologies apart, India is outperformed by higher populated China in terms of poverty reduction. Whereas both countries started in 1948 with similar conditions of rural underdevelopment (particularly constrained by capital needs for expanding irrigated land), they chose unique institutional changes for providing accessible opportunities to the masses of tenants and landless workers. Each path has had distinct distributional implications, plans for employing the gigantic agricultural workforces and patterns of agricultural growth.

India redistributed only 3 per cent of total privately owned land; it managed to intensify cropping, achieve self-sufficiency in food grains and to sustain growth rates in per capita calorie supply. Yet, the concentration of landholdings (Gini coefficient 0.621) and the prevalence of landlessness (nearly 31–35 per cent of the total agricultural labour force) are reflected in the proportion of rural poor in all India (46.1 in 1974; Table 6.8). However, Derez and Sen (1995: Table A.4) show a higher proportion of 56.2 per cent in the same year falling to 44.9 per cent in 1988. Inequality of per capita consumer expenditure in rural areas fell slightly, as the Gini index decreased from 28.5 in 1974 to 27.7 in 1990. Significantly, the real wages of agricultural workers increased at an annual average 2.9 per cent during the same period.

Poverty reduction

The aggregative time trend of poverty prevalence at the all-India level is, however, misleading because of the interstate variation in the scope of land reform. The relatively more aggressive reforms in the state of Kerala fixed the lowest ceiling and abolished tenancy by converting tenants to owners. Table 6.8 shows the available estimates of the poverty prevalence in rural areas for all India and in the state of Kerala, India's most densely populated state. In 1981, Kerala's rural population was 25.4 million with 654 persons per square km compared with 220 in all India. Kerala has also the highest percentage of redistributed land: 17.5 per cent compared with 3 per cent in all India. Kerala is usually presented in the literature as the state that has gone as far as possible to realize an egalitarian agrarian system backed by unionized agricultural workers and progressive political organization.

Table 6.8 Variation in poverty reduction between all India and Kerala, 1956–78

		Prevalence of poverty		
		Rural poor (%)	No. of rural poor (in millions)	Sen's index of poverty
*All India	1956–71	54.1	178.5	0.23
	1973	46.1	208.4	0.17
†State of Kerala	1961/2	503	7.1	0.21
	1977/8	40.9	7.4	0.15

Sources: *Ahluwalia (1985: Table 7.1); †Jose (1983: Table 5.9).

Considering that Kerala is included in the data on all India despite uniformity of measurement and the time periods not being identical, the data suggest a variation in the pace of poverty reduction. Whereas poverty was reduced at the average rate of 4.7 per cent in one decade in all India, the rate was 5.9 per cent in Kerala. The more comprehensive and sensitive Sen index also indicates a smaller poverty gap and more equal distribution of income among the poor in Kerala than in all India. It seems that Kerala's land reform has been a main contributing factor to this differential. There are several other factors, including average size of holdings, agricultural growth in the food sector, pricing policy, higher rates of real wages in agriculture, strong labour unionization in rural areas and the proportionately higher percentage of public expenditure on health and education in the state government of Kerala.[12] These variables have been manifested in different characteristics of rural poverty between Kerala and all India, and documented by Jose (1983). Based on official figures, his findings indicate a higher quality and quantity of life in Kerala's rural areas than all rural India, as shown in Table 6.9.

Iran

The economic system under which land reform was initiated between January 1962 (original law) and January 1963 (annexed law) was based on private property. From the present author's first study of land tenure in Iran in 1955, it emerged that the social organization of its economy determined the distribution of benefits from oil and agriculture in favour of the privileged few, while denying the needs of the many.[13] With over two-thirds of the total population in 1955 living in rural areas, the vast inequality in the distribution of wealth in Iran presented the extreme injustice of a quasi-feudal system.

Pre-1962 agrarian conditions

Furthermore, wealth in rural Iran was based on the private ownership of entire villages and the landlords' control of the most scarce resource, groundwater channels (*qanats*). My 1955 study also revealed that the royal family, together with large landlords and the Islamic establishment (the *waqf* endowments), possessed nearly 80 per cent of the total cultivable land and its irrigation water *qanats*. Rich

Table 6.9 Comparison of quality and quantity of life between Kerala and all India, 1970–81

	Kerala	All India
Life expectancy (years) (1978)	64	52
Infant mortality (per 1,000) (1980)	40	123
Illiteracy rate (percentage of all adults) (1981)	30.8	63.9
Landless as a percentage of total agricultural households (1970–75)	27	31

Source: Jose (1983).

landlords were absentee, living in Tehran, Isfahan or Europe, and their land was mainly cultivated by permanently indebted peasants. During my visit, I estimated the average annual income of these peasants at US$50 per person in 1955, which amounted to less than one-quarter of the national average and one-tenth of the price of a kilogram of Iran's caviar, consumed by the rich in less than an hour. As a result, the distribution of national income by the mid-1950s was highly unequal: the share of the poorest 20 per cent of the total households was only 4 per cent, while that of the richest 5 per cent was 32 per cent.[14]

To grapple with injustice among the majority of the rural people, and without weakening capitalism and antagonizing the landed aristocracy, the Shah decided in 1951 to sell part of his private Crown land and some state-owned land at 800–1,500 rials per hectare. The Shah's land redistribution was managed by a private agency called 'Bank Omran', while most of the beneficiaries from his programme were peasants and even some landlords, city merchants and military personnel. Yet, the urge for a genuine redistributive reform continued.[15] The result was the land reform laws of 1962, which limited ownership to a maximum of one village, with the balance purchased by the government for redistribution to the occupying tenants or sharecroppers. The area of a village ranged from 300 to 600 hectares. Later, the law of 1963 (the annexed law) specified maximum land areas, ranging from 20 hectares in irrigated rice-growing areas to 150 hectares in rainfed wheat-growing areas.

Redistributive consequences of the 1962 law

I have always found it problematic to assess the redistributive effects of this puzzling, complicated programme. During my field visits in 1965, 1970 and 1972, it appeared that official statistics were exaggerated. Likewise, several heads of the departments concerned tended to discredit the achievements of their predecessors and to give inflated figures differing from those of the Central Statistical Office.[16] But reliable sources agreed that the landless workers – estimated at 1 million in 1975 – were excluded from land reform and not all eligible tenants have benefited. Yet, without a violent upheaval (such as the *coups d'état* in Egypt, Iraq, Libya and Syria), nearly half the agricultural households have received title to farming units which they cultivated as tenants. In the meantime, the landlords have retained ownership of the most fertile land in Iran.[17]

Colonel Valian, a confidant of the Shah, brought from the army in 1963 to command the powerful Land Reform Organization and later Minister of Rural Affairs, arranged to pay a total of US$41 million from oil revenues (25 per cent of total compensation) immediately and in cash to all affected landlords. The payment of the rest was guaranteed by the Central Bank. The land reform beneficiaries, in turn, paid the government the purchase price by annual instalments, bearing no interest, but it was disguised in a high administrative cost of 10 per cent. The peasants' estimated gains were represented by the difference between the annual instalment and the share of the output (or annual rent) previously given to the landlord, plus expected increases in yields minus taxes, according to the findings of my 1965 field study of a small sample of 29 land reform beneficiary households in Meshkin Abad village (45 kilometres west of Tehran). With the two improved irrigation *qanats* and easy access to subsidized seeds and fertilizers, the new owners were able to grow two crops annually plus vegetables. From a unit of 7 hectares, each beneficiary paid to the government an annual instalment of 5,300 rials plus 50 rials for a cooperative share, compared with an average of 13,000 rials (my calculated value of the crop share previously given to the landlord). Some families increased their total earnings further by weaving carpets.[18]

The rejoicing of the owners at the rewards they reaped from operating their individual farms was short. In 1967, the Farm Corporations Law was issued, pooling all the lands of individual peasants to establish large-scale and heavily mechanized farms managed by government officials.[19] My visits to three farm corporations in 1970 and 1972 convinced me of the aggressive bureaucratic procedures that were followed, while ignoring the peasants' own preferences.[20]

After Ayatollah Khomeini's revolution in 1979, and based on different interpretations of the Islamic principles regarding property rights, conflicting views have emerged between the pro- and anti-land reform factions. Some mullahs (religious leaders including ayatollahs) in the powerful Council of Guards and the *Majlis* (parliament) condemned land reform as a violation of private property rights, and called for the return to their original owners of cultivated lands expropriated by the 1962 and 1963 laws during the Shah's regime. Ironically, this faction, in coalition with others, was instrumental in the nationalization of privately owned manufacturing industry and major trading and banking enterprises in 1979–82. Another faction took the opposing stand of continuing, with minor modifications, the Shah's land policy in order to promote social justice, considered to be the primary objective of the revolution. The conflicts between pro- and anti-land reform factions within the Khomeini administration lasted a decade (1979–89), until Khomeini himself established the Determination Council in March 1989 to resolve disputed land tenure issues. The outcome was a redistribution of some 450,000 hectares of recovered land, privately owned by landlords who had fled the country, including the royal family, in addition to the distribution of 600,000 hectares of public land.[21]

Morocco

The story of Moroccan land reform experience differs from that of the formerly French-occupied Tunisia in origin and the path followed. The duality of the agrarian structure, associated with the formation of large private estates, started long before the French occupation in 1912. Tracing this origin helps to explain the current concentration of wealth in rural Morocco.

Historical forces behind land accumulation

With their increasing profits from trade between Morocco and Europe, the rich merchants of Fez expanded their wealth in rural areas through holding large tracts of agricultural land in order to influence the central government (*Makhzan*). It seems that the motive behind their land accumulation was to gain political and economic power in the control of cereal supply and trade in rural Morocco. They succeeded, and their powers were reinforced by receiving land grants from the sovereign (*Moulay*), who held the absolute property right of state-owned land.[22] Large areas of 300–600 hectares were granted to influential families from the cities of Fez and Meknes, senior government officials, members of the Moulay family and to Muslim leaders (*Shorafa, Ulamaa* and *Qadi*). Through legal manipulation, the usufruct rights in the granted lands (*Iqta* and *Azib*) were gradually converted into private property (*melk*). Lazarev (1977) gave a detailed account of this process of property rights transfer and listed the names of recipient families, many of whom are presently dominant in Moroccan agriculture and in the national political arena.

In addition, the influence of these families on the bureaucracy enabled them to extort additional areas of cultivable land belonging to small landholders (*paysannerie*) and to register the grabbed lands (*ghasb*) as their private properties. They also took advantage of the sale of state-owned land on favourable terms, following the government's financial crisis caused by the high public spending on the Spanish–Moroccan War of 1860. The economic crisis was deepened by frequent droughts and the outbreak of plague and famine in 1858 and during the 1878–82 period. Numerous hungry and dispossessed small peasants fled their land, seeking food security in towns. In distress, they sold their land and livestock at very low prices (Lazarev 1977: 84–5). The rest of the land, partly growing wheat and barley and partly used for grazing, was held collectively by tribes including the two powerful tribes: the *guich* in the western plains and the *berbers* in the middle Atlas mountains.

Against this background, the French occupation or Protectorate began in 1912. Before that year, the Moroccan agricultural development potentials were carefully studied by several groups of French politicians and technicians. By 1953, there were 4,270 private French settlers owning 728,000 hectares with an average of 200 hectares each. The colonized land was mostly situated in the Casablanca and Rabat regions of the Chaouia and Gharb plains near the Atlantic Coast as well as in the Mediterranean plains. In all, some 6,000 French settlers and foreign companies (both official and private) owned nearly 1 million hectares of fertile and irrigated lands, most of them around the cities of Fez, Meknes, Rabat and Marrakesh (Swearingen 1988: Table 8). More than half the total perennially irrigated land was owned by European settlers who represented only 5 per cent of total landholders. They dispossessed Moroccan Muslims, despite strong opposition from the sovereign *Khalifa Mawlai Abdul Hafed* and the Muslim leaders against foreigners owning agricultural land.[23] As remarked by a French writer who lived in Morocco before independence, the people could only see the transfer of their land to foreigners, nicely disguised in formal judicial phrases (Pascon and Ennaji 1986: 86). The majority of Muslim farmers, numbering about 900,000 families, lived from 6 million rainfed hectares, mostly for grazing, livestock raising and rainfed cultivation (*bour*), which grew cereals under the fallow rotation system.

Half the total number of their holdings were less than 2 hectares in size. Landless workers were about one-quarter of total agricultural households, and another 15 per cent possessed less than 0.5 hectares each, many of whom were also hired labourers. There were also rich Muslim farmers owning more than 50 hectares each, and many of them benefited from irrigation investment during the French administration. As noted earlier, they were already dominating the rural economy before French rule.

It was this agrarian setting that determined employment opportunities and generated conditions of poverty and food insecurity in rural Morocco, particularly in the south. Only 10 per cent of total arable land, mostly in the north and west, was irrigated in 1950, growing not cereals, but the high-value cotton, citrus fruits and vegetables. During the period 1945–50, cereals occupied 60 per cent of total arable land, and wheat and barley were cultivated in 81 per cent of the total cereal area. During the prolonged drought of 1945, the cereal reserve was depleted, and a disastrous famine killed half the total livestock and an unspecified number of poor *fellaheen in* rainfed lands (*Production Yearbook* vols I, IV and V). As documented by Nouvele (1949) and Swearingen (1988: 122), starved and distressed small farmers sold their lands to merchants and fled to the cities. This human disaster led to the issuing of Law No. 5, prohibiting any sale or mortgaging of lands below 7 hectares rainfed and 1.5 hectares irrigated.

The contents of the partial agrarian reform

Like Tunisia, the repossession of foreign-owned land was not of immediate priority in Moroccan post-independence rural development strategy. Instead, priority was given, between 1956 and 1960, to the expropriation of the relatively small area owned by the Moroccans who collaborated with the former French rulers and those who opposed the activities of the Liberation Army. The area affected was nearly 12,000 hectares or 0.2 per cent of the total area of landholdings. Pascon's study of the Haouz region of Marrakesh indicates that only 19 per cent of the area seized was redistributed to tenants who were actually cultivating the lands, while 81 per cent was retained by the state and leased out through public auction (Pascon 1986: Table 3.1)

It was only in September 1963 (8 years after independence) that the decolonization of foreign-owned land was initiated by *al-Dahir al-Sharif* (decree) and was included in the First Development Plan, 1960–64. An area of 250,000 hectares of official colonial lands was gradually acquired by the state through the land reform laws of 1963, 1966 and 1973. However, after the declaration of independence, the sale of privately colonized land was bilaterally arranged between individual foreign owners and Moroccans, including senior government officials and city merchants. The foreigners' unsold privately owned land, together with the area of ex-French official colonization, amounting to 740,000 hectares, was acquired by the state.

According to the 1989 Report of the Ministry of Agriculture and Agrarian Reform (p. 43), only 327,008 hectares was slowly redistributed to 23,600 families between 1966 and 1985. The rest (about 440,000 hectares) was retained by the government and managed as state farms. This means that the number of beneficiaries represented only 1.6 per cent of total agricultural households. The size of distributed units varied according to the productive quality of the land, ranging from 5 hectares of irrigated land to 16–23 hectares of rainfed land per household. Approximately one-quarter of distributed units were in irrigated zones. As followed in Tunisia, the criterion for distribution was a minimum annual gross income per household. On average, the value of the produce from a distributed unit was 4,000 dirhams a year (5 dirhams equalled US$1 at the official exchange rate). This annual income level corresponds to the overall average (not minimum) expenditure per rural household. It was reported that the average level of the sale price was half the market price payable over 30 years in annual instalments.

Equity and production consequences

Thus, by 1985, the state had become the largest single landowner in the country, owning about 440,000 hectares or 6.5 per cent of total cultivable land. In addition, the state owned forest lands and most of the range lands, and held the property rights (*raqaba*) of nearly 1.5 million hectares of tribal lands. Clearly, and despite the uncertain statistics, there was no shortage of productive land at the disposal of the state for redistribution to landless peasants if there had been a firm government commitment to provide greater access to land, and to reduce inequality of income distribution as a priority in rural development strategy. This low priority was given despite the high inequality as well as the high prevalence of rural poverty, estimated at 32 per cent by the World Bank in 1985 (El-Ghonemy 1993: Table 7.4).

In addition, available data on the distribution of *melk* (private) lands (excluding land owned by foreigners, *habous* and tribal lands) show that, at the time of declaring the 1963 and 1973 land reforms, 69 per cent of total agricultural households were landless farmers and owners of less than 2 hectares (Griffin 1976: Table 2.4). Most of them were living in absolute poverty, particularly in the south, where dry farming and pastoralism prevail. At the time of the land reform of 1973, the World Bank estimated that 45 per cent of the rural population were living below the poverty line. However, as Table 6.10 shows, after instituting a series of partial but politically publicized reforms, concentration of land remained high at 0.76 index of inequality (the Gini coefficient). In 1974, 10 per cent of the land area fell into the largest size group of 100 hectares and over, which was held by only 0.2 per cent of total landholders with an average of 278 hectares each. Swearingen (1988: 179–80) reports that, in 1986, of the total of 625,000 hectares of modern irrigated land located in the major irrigation areas, large landowners owned some 500,000 hectares.

The situation improved slightly after the completion of the redistributive programme, as suggested by the results of the 1982/83 survey conducted by the

Table 6.10 Post-land reform distribution of landholdings by size in Morocco, 1974

Size (in hectares)	Number	Percentage	Area (in hectares)	Percentage
< 5	1,079,090	73.6	1,771,900	24.5
5–10	219,790	14.9	1,507,200	20.8
10–20	114,050	7.8	1,525,200	21.0
20–50	43,840	3.0	1,215,300	16.7
50–100	7,720	0.5	512,300	7.0
100 and over	2,520	0.2	699,500	10.0
Total	1,467,000	100.0	7,231,400	100.0
Gini coefficient of land concentration	0.755			

Source: Ministry of Agriculture and Agrarian Reform, *Recensement Agricole, 1973/74*, in Arabic. Percentage errors in the Ministry's table are corrected.

Note
The Gini coefficient is calculated by the present author.

Ministry of Agriculture. Measured in terms of the Gini coefficient, inequality was slightly reduced from 0.76 in 1974 to 0.69 in 1982/83. Thus, land distribution remained grossly unequal, despite the high expectations of the *fellaheen* fostered by government's promise, political unrest (1965–71) and a persistent demand for a wider redistribution of land.[24]

The impact on food production

To understand the implications of this policy for food security, we need to take into account the following agrarian characteristics: (i) the limited scope of the redistributive reforms by which only 2 per cent of total agricultural households were land recipients; (ii) the high degree of land concentration, combined with a high percentage of pure landless households, estimated at 33 per cent of total rural families (Ennaji and Pascon 1988); and (iii) the limited share of agriculture (10–13 per cent) in total institutional credit supply, which benefited about 30 per cent of total landowners, mostly medium and large farmers (*Caisse National de Credit Agricole* and *Banque du Maroc*). Considering that approximately 60 per cent of total cultivable land grows wheat, which constitutes, on average, two-thirds of the consumer's calorie intake, wheat price policy has influenced both land allocation among crops and food security. Apart from the great instability of rainfall, a combination of land tenure arrangements, crop pricing and subsidization policy in the 1970s had an effect on farmers' incentives and food production between 1970 and 1980. The average annual rate of food production growth was very low at only 1 per cent, while population grew at the higher rate of 2.4 per cent; the inevitable result was a fall in food production per head of 1.4 per cent and a sharp rise in total food imports by 59 per cent.

The Philippines

The policy issue of land reform was, and still is, very heatedly debated in the Philippines. Since the 1970s, a myriad of programmes for improving tenancy arrangements and promoting employment and income have been implemented, including land settlement schemes, land consolidation, subsidized agricultural credit, farmer training schemes, integrated area projects and programmes specifically geared to women.[25] But a number of Filipino scholars and non-governmental organizations questioned whether government expenditure in most of these programmes would reach the rural poor, and have forcefully argued for 'a genuine and comprehensive land reform programme'. The call for such a programme became more emphatic after the rise to power of Corazon Aquino, whose pronounced objective was to reduce poverty and the concentration of wealth in agriculture in order to realize social justice principles set out in the 1986 Constitution (Article XIII). Those calling for a massive redistribution of landed property believed that it was necessary to alleviate violent conflicts, social unrest, widespread deprivation in the rural areas, rising landlessness and falling food productivity.

The features of the partial reform, 1954–85

The concern for these structural problems is justified in the light of persistent land concentration (a Gini index of 0.50 in 1960 and 0.53 in 1981) and the rise in landlessness to nearly 40 per cent of total agricultural households. Both these chronic agrarian problems have continued since three earlier 'agrarian reforms', which left the fundamental features of the institutional frameworks of agriculture, the economic powers of landlords, MNCs and moneylenders virtually intact. This half-hearted reform policy had serious consequences. Despite the wide application of technological advances in agriculture in the 1960s and 1970s, productivity per capita of the agricultural workforce declined at an annual rate of 0.8 per cent between 1981 and 1986, and the per capita food productivity of the total population also declined from 2.4 per cent per year (1971–80) to –1.5 per cent per year during 1981–90.

With capital-intensive industrialization representing a large share of GDP of 32 per cent but employing only 16 per cent of the total labour force in 1986, the percentage increase in new entrants to the agricultural labour force was as high as 1.6 per cent per year between 1970 and 1990. Apart from falling productivity, the results were increasing fragmentation of smallholdings, increasing landlessness, widening inequalities, a rising ratio of agricultural population to cultivable land from 2.6 persons per hectare in 1960 to 3.4 in 1990 and an increasing frequency of land seizure (called land takeovers or invasion by the landless poor).[26]

Income effects

The first two land reform laws of 1954 and 1963 were attempts to improve tenancy arrangements in sharecropping, to lower rents and even to abolish tenancy altogether. But unenforceable laws are not laws. The third land reform of 1972 gave the tenants cultivating rice and corn a choice of options. The tenant could own the piece of land he tilled against payment through transferring title from landlords whose ownership exceeded 7 hectares or could rent at 25 per cent of the average net value of output, while being granted higher tenancy security. The clumsy bureaucratic procedures and the influence of the landlords significantly slowed the pace of implementation. The end-result was that only about a quarter of the total number of tenants in rice and corn land were able to purchase land from their landlords or to hold title once payment was completed after 15 years.[27]

As in other PLR countries, landlords retained the most productive portion of their land together with substantial capital assets, leaving the new owners to purchase the less productive and decapitalized land. New owners did, however, gain an immediate income transfer from the amortized annual payment of land price and land tax – payments that were lower than the former rent of 50 per cent of the net value of harvest. The income gain was more widespread among the tenants who continued as leaseholders and who paid 25 per cent of the harvest in addition to their prescribed security of tenure under written contract. However, these provisions were not uniformly enforced. Using the results of microstudies, Mangahas (1985: 235) reports that, in rainfed areas, tenants continued to pay more than the 25 per cent legal limit. These studies indicate also that 58 per cent of hired workers in sugar plantations in the Negros Occidental province received wages less than the minimum required by the 1975 Wage Commission Rules.

The 1972 land reform was partial because it was limited to rented lands growing rice and corn. It excluded cash crops (coconut, sugar cane, banana, pineapples, coffee, tobacco, rubber and cattle ranches) and the owner-operated farms irrespective of their size. The cash-cropped area included 3.1 million hectares (mostly irrigated land), 170,000 tenant households and 1.1 million landless workers (Mangahas 1985: 219). In most of this cash-cropped sector, production relations and market structure were dominated by large landlords and MNCs. Thus, a duality in the rural economy was created between the beneficiaries and the landlords whose ownership exceeded 7 hectares within the reform sector, and between that sector and the large cash-cropping and export-orientated sector. In both sectors, the market power of the landlords, traders and moneylenders has continued in terms of their share in the rural banks' stock capital, the acquisition of the means of production and in their marketing of farm produce. Also, the vast majority of wage-dependent landless workers have continued to live in poverty (Alex and Almeda 1980; Ledesma 1982). With inflation rising at an annual average rate of 15–20 per cent, real wage rates of the landless workers declined (wage index of 100 in 1972 fell to 69 in 1980).

Poverty prevalence

The microstudies cited above indicate that the annual earnings of landless workers, averaging P2,000, are less than half the minimum income fixed by the World Bank as the poverty line. In the size distribution of real income, the share of the bottom 30 per cent of the rural population fell by 20 per cent between 1970 and 1980 (Sobhan 1983: 9, Table 11). Despite continuing controversy in the Philippines about the mathematics of counting the poor, there is consensus that the numbers have risen, particularly among the landless labourers in agriculture. The National Nutrition Survey, the research of the Philippine Development Academy and the studies of the World Bank in 1980 indicate an upward trend in the number of poor people irrespective of conflicting views on the criteria for the cut-off point (poverty line).[28] We used the World Bank estimates for 1965 and 1980/1, during which period the land reform was implemented. These estimates show a small proportionate decline at the rate of 1.5 per cent per decade, but a rise in the absolute numbers of the poor by nearly 3 million persons in the countryside. The rise would be much greater if other estimates were used. As in the case of Egypt, I shall continue in Chapter 7 to outline the equity changes arising from the adoption of the market liberalization policy.

Syria

Despite being the last case study of PLR, this is the most recently examined (my field study was in May and June 2004). Interestingly, the Syrian constitution of 1950 in its Article 22 and the declared objectives of the 1958 Land Reform Law as well as the key objectives stated in the first *Five Year Development Plan* of 1961–65 have clearly stressed equity and production incentives of reforming land tenure arrangements and their intrinsic connection with rural poverty reduction and abolition of exploitation (*Istighlal*), the meaning of which has already been discussed in Chapter 2.

The redistributive contents of land reforms, 1958 and 1973

Accordingly, the 1958 and 1973 redistributive reform laws established two different ceilings on private landownership in irrigated and rainfed areas, above which land was expropriated against compensation payment for its redistribution among peasants. The small family farm was established at 8 hectares of irrigated land and 40 hectares in rainfed areas. In addition, the shares of the owner and tenant/sharecropper in leased areas were specified, while protecting the tenants from eviction, and the conditions for hiring wage workers were laid down in formal contracts to be monitored by the Ministry of Labour and Social Affairs. By 1969, an area of 1,513,812 hectares had been expropriated, of which only 695,843 hectares was redistributed, representing 27 and 12 per cent of total cultivable land respectively (Arodki 1972: Table 5). An additional area of privately owned land amounting to 42,700 hectares was expropriated by the land reform law of 1980.

The balance of unredistributed land representing such a high proportion of 60 per cent of the total expropriated area has, since 1974, been retained by the government (*Amlak Al Dawla*) and used partly as state farms and partly leased out to tenants at a fairly fixed rental value. Starting in 2000, sold state farms have not been allotted to landless *fellaheen*, who are the intended beneficiaries of the land reforms, but instead managed jointly with the private sector as joint-venture companies of large-scale agricultural production and processing 'projects', employing intensive capital and not intensive labour.

Halting the redistribution of a substantial area of expropriated land has diminished the peasants' opportunity to have secure access to land as private landowners, which was promised in the 1958 land reform. Consequently, the average size of landholdings has declined since 1970 from 11.8 hectares to 7.6 hectares in 1994. In addition, the large holdings in the size category of 50 hectares and over, representing only 2 per cent of the total landholders, have maintained a high share of the total area at 23.5 per cent in 1994 with a tendency towards land concentration in the category of 10–50 hectares. As shown in Table 6.11, the land area share of this category of landholders – who represented nearly one-fifth of the total – had increased to almost half the total area in 1994, according to the results of the 1970 *Census of Agriculture* compared with the last available census of 1994 (at the time of my field study in May–June 2004, the results of the 2004 census were not available).

The effects on production, landlessness and rural poverty

Being partial in its redistributive scope, Syria's PLR has had limited effects. In fact, much of the optimism about the productive absorption of the landless rural poor by land reform has been on the wane for several reasons. The first is the successive subdivision of properties among household members (about 8–10 persons per household on average) through inheritance arrangements. The second is that many poor agricultural households were left out by the land reform implementation and remained small tenants, sharecroppers and landless wage workers. The numbers and proportion of these disadvantaged groups have increased steadily, reaching 34.7 per cent of the total agricultural households covered by a sample survey of eight villages in Muhafazat Idlib, Hamah and Al-Hasaka in 2000 (Forni 2001: Table A3.1) where poverty was prevalent.

Available official statistics on agricultural households *without* land provided by the 1994 Census of Agriculture tell us that their numbers have almost doubled since 1981 (11,224 in 1981 and 22,860 in 1994), and are concentrated in the Muhafazat of Al-Rekkah, Deir-ez-Zor and Al-Hassaka, where rural poverty is con-

Table 6.11 Size distribution of agricultural landholdings, 1970 and 1994, in Syria

Size of holdings (hectares)	1970		1994	
	Number (%)	Area (%)	Number (%)	Area (%)
< 2	32.2	2.8	35.7	4.0
2–10	42.9	21.6	41.6	23.3
10–50	22.1	46.1	20.7	49.7
50 and over	2.6	29.2	2.0	23.0
Total	100.0	100.0	100.0	100.0

Source: The results of Agricultural Censuses 1970 and 1994 taken from El-Zoobi (1984: Table 39) and *The State of Agriculture in Syria*, FAO (2003: Table 11.5) respectively.

centrated according to the results of 2003/04 Household Income and Expenditure Survey (HIES). In addition, there were 110,000 tenants leasing or sharecropping farm units of less than 2 hectares each (i.e. near landless peasants) whose well-being suffers greatly in years of drought.

With regard to output growth, we should note that land reform was an integral policy instrument in a broad equitable rural development strategy implemented during 1960–80. Apart from the PLR, the strategy included: (i) the upsurge in public investment in three giant projects for labour-intensive irrigation and land development of nearly 740,000 hectares (Sud al-Forat, Al-Roston Dam and Asharna-Al-Ghab project; and (ii) a pro-farmers' price system including prices of production inputs and subsidies, combined with a substantial investment in human capital. Thereupon, notable well-being achievements were realized between 1960 and 1980; life expectancy increased by 28 per cent, and child death rates and adult illiteracy rates decreased by 76 and 40 per cent respectively.

Accordingly, agricultural output grew fast at an average annual rate of 4.4 per cent in 1960–70 and much faster to 8.6 per cent in 1970–80. During the period 1960–80, the irrigated area was doubled and food production per head was maintained above the average level (1960–65 = 100), despite rainfall instability.[29] This notable agricultural performance was achieved in spite of rapid population growth at an average annual rate of 3.5 per cent and much higher in rural areas, constituting 50.3 per cent of the total population in 1980–2001, and reaching over 60 per cent in six provinces or Muhafazat.[30]

Using the results of the last two available HIES surveys of 1996 and 2004, we find that the prevalence of poverty has increased in the rural areas of the north-eastern Muhafazat (according to the lower poverty line established by the government). Forty-six per cent of the total poor lived in these Muhafazat in 2004, and most of them are landless agricultural workers (El-Ghonemy 2005: 59–61).

Summary

Given the political economy of each country, our six case studies suggest that each country leadership designed the PLR as a compromise to partially meet the aspirations of the poor peasants and, at the same time, not to damage the interests of landlords/capitalists, whose long-term support was needed by the reformers for preserving political and economic stability. This policy motive is manifested in both the deliberate exemptions made and the promulgation of the piecemeal acts. For example, Egypt, Morocco, the Philippines and Syria promulgated PLR in three or four laws each, over two decades, while India issued numerous laws between 1952 and 1979, yet it distributed nearly 3 per cent of total agricultural land to only 4 per cent of total agricultural households in 1980.

The case studies also suggest the momentum of PLR as an integral part of an equitable rural development strategy, comprising the construction of giant irrigation schemes for the purpose of a rapid expansion of irrigated land, pro-poor pricing policy, fixing minimum daily wages, non-farm employment expansion, combined with fast-growing remittance transfer to rural areas, population fertility control and, importantly, public investment in human capital. The countries have made remarkable progress in literacy and life expectancy, etc.

It is difficult to ascertain whether the combination of PLR and these poverty-reducing measures was intentionally conceived by the country leadership for effective land reform or if these measures were separate policy instruments for participatory rural development conceived within an overall national development framework. From my reading of PLR objectives and personal discussions with senior officials in these countries, it is apparent that there was a combination of post-1950 ideological shift towards interventionist policy for greater equity and peasants giving vent to their long frustration from injustice. Post-land reform achievement of aggregate agricultural growth and poverty reduction in rural areas cannot, therefore, be solely attributed to the PLR provision of secure access to land for 3–10 per cent of total agricultural households over three decades, but importantly to the rural development impulse.

7 Market liberalization

Market-based land reform

This chapter is mainly about post-1990 market liberalization concerning land access. The discussion consists of four major sections. It begins with an attempt to understand the origin of the post-1980s policy shift away from direct government action to amend the land market failure by way of state-mandated redistributive land reform, which had been a key public measure for poverty and inequality reduction in rural areas since the late 1940s. The second section presents the market-orientated or assisted or negotiated land reform introduced as an integral part of the neoliberal economic reforms. We have already presented, in Chapter 2, the conceptual elements of neoliberalism and the main features of the land market approach to poverty reduction via voluntary land property rights transfer between a willing buyer and a willing seller at market price, supported by external financial assistance. The third section examines the closely related privatization of the historically long-established communally owned customary land in Latin America and many African countries. The last section identifies the impact of market liberalization on existing state-mandated redistributive land reform in a selected sample of six countries: China, Egypt, Nicaragua, Honduras, the Philippines and Russia. Being the mother of communist Marxist ideology on reforming the agrarian structure, Russia is included in order to learn some useful lessons from its recently introduced market mechanism.

Understanding the origin of the policy shift

Although land reform is primarily the responsibility of the sovereign governments of developing countries (LDCs), donor countries and international financing agencies (IAs) have a powerful influence on their policy-makers. They do play an important catalytic role in shaping LDCs' policies on resource use and income distribution, despite the relatively low share of aid in total investment. They rely mostly on the use of both linguistic devices and the financial instruments of power.

The central issue addressed here is stated in the form of a hypothesis: 'the shift in the content of policy prescriptions provided in the foreign assistance to LDCs corresponds to shifts in the internal ideologies of the prescribers and not necessarily to changing conditions of rural poverty in recipient LDCs'. The assumptions on which the exploration of this proposition is based are as follows:

1 The content of the donors' and IAs' policy prescriptions for how to approach LDCs' rural development problems and what should or should not be done is a function of the former's set of ideologies and internal politics.
2 The technical functionaries (including economists) employed by the donors and IAs are intellectually conditioned by their employers'/institutions' ideologies even if, in the functionary's set of beliefs, the policy prescriptions are irrelevant.
3 The test of the donors' and IAs' ideologies is in their actions and the resulting consequences.
4 If the policy prescriptions ignore the LDCs' central development problems of persistent poverty, increased undernourishment and gross inequalities in rural areas, the prescriptions are either irrelevant or pursuing the vested interests of the prescribers.

Perhaps no other policy issue is more susceptible to this ideological complex than land reform and related income and rural power redistribution. Furthermore, considering that the term ideology is ambiguous and does not possess a meaning on which all its users agree, and because I have little competence in international politics, this discussion will be limited to the identification of the elements in the shift in international development assistance in respect of land reform and poverty alleviation over the approximate period 1950–86. To make the discussion manageable, it shall be confined to three IAs considered to be very powerful in the world today: the United States government (USA), the World Bank and the International Monetary Fund (IMF), which, as I mentioned in Chapter 2, are collectively given the term 'the Washington consensus' by the 2001 Nobel Laureate Joseph Stiglitz.

The shift in the US stance on land reform

After the Second World War, US foreign development assistance vigorously pursued land reform as a major redistributive and stabilizing policy. Past experience suggests three possible explanations within capitalist ideology for using this strategy. The first is moral: to satisfy the American ideology founded in the early nineteenth century by Thomas Jefferson and Abraham Lincoln, according to which the entitlement to holding private agricultural land is considered a prerequisite to democracy and market economy. This conception was instituted early in 1862 by the Homestead Act (Farmer-owned Family-size Farm). The second motive is political: to counter potential communist movements among the desperately poor peasants in LDCs. The third is economic: to secure rural stability conducive to American multinational private investment in LDCs' agriculture and to expand their internal market demand for American-manufactured goods. This is because the growth of production and the potential rise in the incomes of land reform beneficiaries are anticipated to lead to an increase in their effective demand.

If my interpretation is correct, we can understand the vigorous intervention of the USA in instituting land reform programmes immediately after the Second World War in Japan, Taiwan and South Korea, and in their provision of financial support for implementing land reform in southern Italy as part of the Marshall Plan for Europe. This successful experience in the late 1940s and early 1950s encouraged the United States to pursue its pro-land reform policy and to extend its assistance to LDCs. To integrate this ideology into the federal government functions, President Truman entrusted its implementation in January 1949 to the newly established Technical Cooperation Administration (TCA) in the State Department. As one of its first acts, the TCA held the first World Conference on Land Tenure (Madison, Wisconsin, 1951, when I was a graduate student in the USA). The conference was instrumental in disseminating subtle ideas on land reform based on professional reasoning and empirical evidence.

At the conference, the senior officials of the US government stressed the elements of 'genuine land reform': redistribution of land; security of tenure and clear titles; access to credit and the 'development of farm cooperatives for cultivating, marketing and processing agricultural products'. This conception was presented by a resolution to both the United Nations Social and Economic Council and its General Assembly in 1951, and to which the US government committed itself 'to encourage and assist countries' (Parsons *et al.* 1956: 26–37). This significant action by the US government was followed by the submission and adoption of a resolution by the FAO Conference in November 1951, which mandated FAO to accord land reform a high priority in its programme of work.[1]

However, where the land reform programmes were designed and intended to expropriate large foreign-owned estates including those of American MNCs, US support was absent, as occurred in Guatemala and the Philippines in the 1950s (for a detailed account, see Olson 1974). Within this framework of inconsistency between ideals and realities, a new initiative was promulgated by President Kennedy in 1961 when he immediately stated his ideals in his inaugural address: 'If a free society cannot help the many who are poor, it cannot save the few who are rich [. . .] our pledge is to assist free governments in casting off the chains of poverty'. The concern of the Kennedy administration was primarily focused on the implications of high land concentrations in Latin America combined with absentee landlords and increasing numbers of poor landless workers who had virtually no chance of acquiring land on their own, through the defective market mechanism. There was also a fear of the potential response of peasants' movements in some Latin American countries following Fidel Castro's successful revolution in Cuba in 1959 and its Soviet-backed land reform policy. This concern was expressed in President Kennedy's words:

> There is no place in democratic life for institutions which benefit the few while denying the needs of the many, even though the elimination of such institutions may require far reaching and difficult changes such as land re-

form and tax reform and a vastly increased emphasis on education, health and housing.[2]

The events that followed suggest that his statement was not merely political rhetoric, but a genuine commitment reflected in four initiatives taken by the US government in the early 1960s:

1 The creation of the Alliance for Progress programme, in which land reform, combined with social services, was central, and for which the US Congress allocated US$500 million.
2 The establishment of the Inter-American Development Bank (IDB), in which the USA's contribution amounted to 45 per cent of the IDB's capital on the proviso that the funds were to be used for social progress, including land reforms (*reforma agraria*).
3 The establishment of the Land Tenure Center at the University of Wisconsin for training, research and for furnishing policy advice to LDCs' governments.
4 The issue of the Foreign Assistance Act 1961 and the creation of the Agency for International Development (AID) to implement the programme. Sections 102 and 103 of the Act are directed to land reform, and it is useful to quote a few lines, which indicate the perception of the US government of the time as follows:

> The principal purpose of bilateral development assistance is to help the poor majority of people in developing countries to participate in a process of equitable growth through productive work and to influence the decisions that shape their lives, with the goal to increase their incomes [. . .] the establishment of more equitable and more secure land tenure arrangements is one means by which the productivity and income of the rural poor will be increased.

This powerful reasoning was articulated by the American Secretary of Agriculture at the USAID Spring Review on Land Reform, Washington, DC, June 1970, and the World Conference on Agrarian Reform and Rural Development, Rome, 1979 (the present author participated in both conferences). In 1980, the Republican Party came to power and, under the Reagan administration, the official stance shifted to the opposite pole. Compare the preceding expressions of pro-land reform policy with the new official direction in foreign aid published in 1986 under the title *Policy Determination on Land Tenure*. The dramatic shift is apparent: no support to government intervention in private land redistribution, but the distribution of *public lands* in settlement schemes and cadastral surveys to be financially aided; land to be purchased in the open market with American assistance if needed; and the supply of inputs for production must be done through the private sector if provided by American aid, etc. This new perception was reiterated in November 1987 when the American Delegation to the FAO conference held in Rome insisted that the market mechanism and *not* the government-mandated redistributive land reform should realize equal distribution of land. In addition to this US internal political shift in 1980, we suggest that the collapse of the Soviet Union in the early 1990s, combined with the softening of Russia's and China's habitual aggressive stance behind state-administrated land reform, was associated with their international cooperation with the USA on subjects of mutual interest.

Exploring the World Bank's shifting policy

In the domain of international development assistance and lending, the World Bank is certainly a powerful institution, influencing the thinking of policy-makers, the participants in international development debate and senior technocrats in many LDCs. As a multilateral assistance agency in the United Nations system, the World Bank has a comparative advantage in collecting development data and in undertaking field research on fundamental policy issues. This advantage is reflected in the World Bank's intellectual contribution to the understanding of the relationships between economic growth, income distribution and poverty prevalence. Under the leadership of McNamara, a close associate of the late President Kennedy, the World Bank accorded high priority to agriculture in general, and land reform and rural poverty in particular, between 1972 and 1980. In my view, this was the golden age, so to speak, of these rural development policy issues, both intellectually and operationally. In its guidelines for lending and granting assistance to developing countries, the World Bank stated 'In countries where increased productivity can effectively be achieved *only* subsequent to land reform, the Bank *will not* support projects which do not include land reform' (World Bank 1975: 11). In the 1970s, nearly 55 per cent of the total World Bank lending to agriculture went to anti-poverty-focused rural development projects. This does not mean that all funds invested in these projects reached the poor and directly benefited the landless workers.[3]

As US foreign policy on aid to LDCs shifted, so did the World Bank's by the turn of the 1980s. Its development perception had suddenly changed, and its rural poverty-related lending and assistance work had declined (see Chapter 1, section entitled 'Widening gap between international commitment to reduce poverty and the realities'). The World Bank's new land reform policy prescription is based on a market mechanism freed from government intervention and its central planning for development. Its scope is dependent on linkages between land and credit markets (i.e. the collaterality of land after its formal titling for an efficiency-enhancing land market according to the World Bank's reasoning). This is intended to ensure that resources are used more efficiently by the private sector, leaving the pattern of land and the rural power structure unchanged. 'Unemployment and poverty are to be alleviated by creating a policy environment which will encourage foreign and domestic private investment and makes markets and incentives work' (*World Development Report* 1986: 43). Yes, markets and incentives can work in LDCs, but under what conditions of rural institutions and social organization? Where the distribution of landownership and opportunities is highly skewed, the market works for the benefit of traders, large and medium farmers and MNCs, while most probably harming the interests of poor peasants and landless workers.

In a recent statement, two senior officials of the World Bank articulate the key principles of its current land policy after making a careful assessment of the 1975 *Land Reform Policy Paper* in the light of a better understanding of country experiences and international political changes (Deininger and Binswanger 2002). Briefly, the conceptual foundations are: efficiency of owner-operated family farms and not collective farms; improved functioning of the land market for an efficiency-enhancing transfer of land from less to more productive users based on strong linkages between land and financial markets; and land redistribution with private sector support, whereby the land rental market is more advantageous than the land sales market (Deininger and Binswanger 2002: 411–17).

Could worsening LDCs' foreign debts, terms of trade and balance of payments in the late 1980s and early 1990s justify the sharp turn in the IMF–World Bank policy? Although increased in scale, they are longstanding problems, and it is necessary to solve them. Yet they cannot be solved solely in terms of a financial medicine by the IMF. The problems represent a long-term structural imbalance, which should be resolved within a long-term perspective broad enough to realize sustained rural *development* and not only higher rates of exports and economic growth. Given the realities of agrarian conditions in LDCs, the World Bank's new policy of market liberalization and structural adjustment is likely to exacerbate the already skewed distribution of income and power in favour of the medium and large farmers closely integrated with the international market via exports and the multinationals.

At the time of the land reform policy shift in the late 1980s, the influence of the Group of Seven (Canada, West Germany, France, Italy, Japan, the UK and the USA) is obvious (lately, Russia has joined the group, referred to as the G8), considering their control of 49 per cent of the voting power and 51 per cent of the World Bank's capital stock. The USA holds the greatest share of total capital stock and of total voting power in the World Bank, entitling her to hold senior posts, including the presidency. Yet, the LDCs' problems were compounded in the 1980s by the policies of the members of this group themselves. These policies included offering LDCs excessive lending facilities in the 1970s, followed by an unprecedented rise in interest rates in real terms and an overappreciation of the exchange value of the US dollar. Concurrently, demand for LDCs' primary agricultural products was reduced both in volume and in prices, and a slowdown in the OECD member countries' annual economic growth rates occurred from an average of 4.7 per cent in 1965–73 (before the sharp rise in oil prices in 1973) to 2.8 per cent in 1974–80 and to 2.5 per cent in 1986. This slowdown led to rising unemployment in their economies. Between 1980 and 1985, industrialized countries also reduced their annual rate of concessional external assistance to LDCs. As creditors and importers of primary products, they were also unwilling to open their markets to goods produced by their poor debtors from LDCs.

Historically, and as indicated earlier in Chapter 1, it is unfortunate that the land reform policy shift away from anti-poverty redistributive land reform has taken place shortly after virtually all developing countries had officially committed themselves – in July 1979 at the World Conference on Agrarian Reform and Rural Development – to take action for the realization of an 'equitable distribution of land'. At that conference, on whose organizing staff the present author was an active member, their governments adopted a Declaration of Principles, a Programme of Action and a Resolution in which they agreed in paragraph 8 of the Declaration that, 'the sustained improvement of rural areas requires fuller and more equitable access to land, water and other natural resources; widespread sharing of economic and political power'. They also agreed to impose ceilings on the size of private holdings and to 'implement redistribution with speed and determination' [FAO 1979: paras I, A (iii) and (vi) and II, A (i) and (iv)].

To summarize this section on understanding the policy shift, the discussion suggests the following:

1   State-administered redistributive land reform as a policy issue in rural development has been a victim of changes in the operational ideologies of the major industrialized countries and international institutions entrusted with development aid to poor countries.
2   The shift cannot be attributed to complacency about improvements in LDCs' conditions of poverty. Instead, there has been worsening of poverty and continuing horrendous proportions of malnutrition, landlessness and declining food productivity, documented in Chapters 1 and 3.
3   The stated hypothesis in the introduction to this chapter tends to be confirmed.
4   The identified shifts and their proclaimed reasonings are likely to confuse students of development studies and disturb LDCs' development progress towards poverty alleviation.

On the marketability of land

Lastly, but importantly, the land market approach depends on the marketability of land. Throughout the recorded history of many of the now LDCs, land has been viewed as a secure form of holding wealth and gaining social and political advantage. Unlike other productive assets, land held individually or communally for a long time is almost sacred and, except in distress sale situations, it is preserved as a family or tribal heritable bond, which is immobile and a non-marketable family asset.[4] Thus, in the land market policy prescription and analysis, it is absurd to view land in a narrow economic sense as a commodity or a factor of production, and to analyse the land market as, for example, the fertilizer market.

This principle was established nearly half a century ago by the 1979 Nobel Laureate Arthur Lewis (1955: 91). In his seminal work, 'Freedom of Market' in *The Theory of Economic Growth*, Lewis says:

> There must be access to land [. . .] and clarification of title is a necessary step in economic growth [. . .] Nevertheless, ownership of land is frequently tied up with pride of family, and the latter may cause people to be unwilling to part with land which has been in their family for generations [. . .] Such

considerations are probably most powerful in countries where land is very unequally distributed.

Lewis (1955: 91)

Countries' experiences

We begin with the countries that adopted this approach around 1995 as pilot schemes initiated and generously supported by the World Bank: Brazil, Colombia and South Africa. Their rural economy suffers from persistently high rural poverty (Brazil 40 per cent, Colombia 45 per cent and South Africa 60 per cent) and high land concentration in terms of the Gini index: Brazil 0.91 and Colombia 0.80 (that of South Africa is not available) (see sources in El-Ghonemy 1999: Table 2; and Chapter 3, Table 3.5, of the present volume).

Brazil

Despite the existing high prevalence of poverty in rural areas and the very high degree of land concentration, and in spite of the increasing political importance of land reform, a Brazilian scholar has recently written, 'Brazil has not undertaken, is not undertaking nor will be undertaking a real agrarian reform' (da Veiga 2003: 59).

In addition to land settlement schemes (mostly squattered settlements in the north) that have, since the 1960s, been managed centrally by the federal government, the World Bank-induced negotiated land reform was initiated in 1995 in the north-east as a pilot scheme, following studies supported by eight international agencies.[5] Accordingly, the government appointed a Federal Minister for Agrarian Reform and established an Agrarian Studies Center and a special agricultural credit programme (PROCERA). The targeted household beneficiaries, i.e. possible candidates, were estimated at 2.5 million (Deininger and Gonzalez 2002: 335). Credit assistance being an essential component of the programme, each land purchaser received an amount of US$4,500–7,500 at highly subsidized 70 per cent credit repayable upon reaching the 'emancipation stage', meaning the desired production ability that qualifies the beneficiary for formal land titling. Besides the decentralized management with local community participation, the scheme is monitored by a committee of university professors and government representatives, with expectation for its expansion at the national level.

Available information indicates slow progress, as only 10,000 beneficiary families had purchased land by the year 2000. Evidence also shows that landlords sold low-quality land and that repayment of loans is difficult. This very slow progress should be seen against the estimated target of 2.5 million beneficiaries, representing nearly one-quarter of total landless and near-landless households and an average PROCERA credit of US$4,864 per household in the north-east, where 60 per cent of the rural population live in poverty. This region-specific average is much higher than the national credit average of US§§2,408 plus 7 per cent administrative costs (Wolford 2002: 304).

Colombia

Unlike Brazil, 75 per cent of the land suitable for crop cultivation was devoted to extensive livestock grazing in 1995, and druglords have purchased land as a form of money laundering that created widespread violence in rural areas. To reduce violence as well as the very high degree of inequality in land distribution (0.80 Gini index) and the nearly 45 per cent rural poverty level, a land reform law (No. 160 of 1994) was issued to provide a legal framework and mechanism for market-based land reform. Similar to the World Bank-induced programme in Brazil, this scheme was implemented in pilot areas, and potential buyers of land are lent 70 per cent of the land sale price supported by the World Bank and the International Fund for Agricultural Development (IFAD); the balance (30 per cent) is to be obtained at market rates. Model farm projects are formulated to help beneficiaries to intensify both production and family labour in their farm units of 15 hectares, on average. In target areas, non-government organizations (NGOs) help to locate land for sale, which is mostly underutilized and belongs to large landowners.

Farm plans drawn up by the pilot schemes emphasize production viability and the major role of local universities and local government in *municipios* suitable for small farm cultivation. In practice, there was credit market rationing and imperfect information about both potential land sellers and land sale prices. Available information also indicates the prevalence of bureaucratic procedures, combined with corruption within local offices of the government implementing agency (INCORA). The results of a sample of 220 household beneficiaries conducted in 1997 are summarized by Deininger and Gonzalez (2002: Table 13.1) showing: high cost per beneficiary household (US$13,000); only 20 per cent of all surveyed households knew about the land market procedure; and the complicated legal arrangements to soften Law No. 160 of 1996 proved ineffective (Deininger 2002: 332). According to a study prepared by CEPAL of the UN, the programme has had very limited success, owing to the high prices and violent coercion from landlords and narcotics dealers, as well as the refusal of willing buyers to purchase land in any locality and a cumbersome bureaucracy. The study also found that most of the land buyers are urban and that transaction costs are prohibitive for small peasants. In addition, the findings show that 'transfers of property rights through the market have virtually failed to shift land from one (rich) group to poor peasants' (see Vogelgesang 1996). This remark is manifested in the continued high inequality of land at the Gini index of 0.80, according to the results of the agricultural census of 2001, and a household income/consumption distribution survey which showed a high inequality Gini index of 0.5 in 1999.

South Africa

Unlike Brazil and Colombia, the historical experience of South Africa makes it a special case indeed.[6] Both the agrarian structure and the entire social order were a striking manifestation of absolute injustice. They were shaped on racial grounds, beginning with the long colonial rule of the Dutch and the British, and formalized by apartheid in 1948. When constitutional reforms were instituted in 1995, the minority white population, representing nearly one-tenth of the total, owned most of the agricultural land (83 per cent). In contrast, native Africans, representing 77 per cent of the country's total population of 41 million, accounted for 61 per cent of all the poor, including 31 per cent of rural households who were landless and with no grazing rights (Government of South Africa 1995: 81, Tables 1 and 2). Vast inequalities of income and opportunities are evident from the 1993 data on the distribution of family consumption, which show that the share of the lowest one-fifth of all families in total income was only 3.3 per cent of total consumption, while that of the top one-tenth in terms of income was almost half the total consumption (World Bank 1997: Table 5)

It is in this context that the present land policy has been instituted as part of the 1995 Reconstruction and Development Programme. Its three components reflect the main land tenure defects to be redressed and the adopted post-apartheid course of action. These components are:[7]

1 Land redistribution by way of market-based property title transfers between willing buyers and willing sellers, with government and several donors' financial support; the potential beneficiaries are estimated at nearly 1 million landless workers and 200,000 tenants – the plan was to redistribute nearly one-third of the targeted area, totalling 30 million hectares, i.e. nearly 8 million hectares for redistribution during the first 5 years of the programme (until the year 2000); the targeted beneficiary families during this period totalled 3 million (nearly 600,000 beneficiaries per year on average).
2 Land restitution for the black Africans who were dispossessed after 1910–13 without compensation, and who were moved out by whites and concentrated in designated 'homelands'.
3 Land tenure security for strengthening tenants' lease rights and the protection of customary land tenure arrangements, with emphasis on the rights of women.

The programme has been implemented with a strong political commitment and partnership between the Ministry of Land Affairs and NGOs. However, implementation has been slow, owing partly to the still rigid racial structures in the government administration, and partly to exorbitant land prices, which are negotiated from very unequal bargaining positions in face-to-face encounters between the many poor peasants willing to buy and the few powerful landowners.

As in Brazil and Colombia, the programme's implementation focused on beneficiaries' groups (200 households in each group, on average) and productive farming in pilot schemes. Cross *et al.* (1996), in their study of land reform pilot project areas in the province of KwaZulu Natal, identified the lack of coordination between the several agencies for the delivery of production services and the tribal chiefs' interference as an exhibition of their traditional power in their localities. In his analysis of the results of a survey of 1,168 household beneficiaries in 87 land reform projects conducted in late 1999, Deininger (2002: 342 in de Janvry 2002) reports that 'not a single one of the 13 projects sampled in KwaZula Natal makes a profit'. The results also show some features of successful projects. These features are: cash contribution of beneficiaries; projects having access to group credit and managed by an appointed manager; and a higher level of education among the beneficiaries. Although it seems easy to criticize this comprehensive programme, several factors are assuredly hopeful: government commitment; support from international aid agencies and OECD member donors;[8] and transparency in the management and work of the Monitoring and Evaluation Unit of the Ministry of Land Affairs.

Yet, there are signs of dissatisfaction with the path chosen. For example, 73 South African NGOs declared in their Charter on Land and Food Security that 'if land reform was left entirely to the market, little if any reform would take place', and 'land reform policy must be driven by the principles of social justice and basic needs as opposed to market forces'. And the National Land Committee has stated that 'the market is not a solution for a fair land redistribution after the apartheid [. . .] markets are never truly free'. The Committee proposed 'a more interventionist role for the government to achieve a thorough and speedy redistribution of land' (IFAD/World Bank/FAO 1997: 43). Likewise, in November 2001, the National Land Tenure Conference called on the government 'to accelerate land redistribution' because many landless workers with insecure employment are increasing their demand for secure land access or land property transfers. The Conference emphasized 'the need for the government to decentralize land administration'. The delegates felt that 'Government has been very slow in implementing the land policy of restitution' (Roth 2003: 3–5).

A South African analyst reports that, by 2002, only a little under 1 per cent of the targeted area for redistribution was realized, and adds 'The assumption that the market could deliver a significant amount of land to poor rural, black people was flawed from the start [. . .] the concept of the market as a mechanism for redistribution, remains abstract' (Levin 2002: 172). This very slow progress was also remarked in 2000 based on a study of 3 years' implementation, during which only 20,000 beneficiaries possessed 200,000 hectares, a ratio of less than 1 per cent of the targeted 18 million hectares and 1.8 million beneficiaries over 3 years (F. Zimmerman, 2000). One can also say that it is absurd that the South African natives have to become indebted in order to purchase their own land that was grabbed by colonial Europeans since 1913 during the apartheid regime.

In a conference on agrarian reform organized by the German Foundation for International Development, held in Bonn in March 2001 and attended by 100 delegates from different countries, including the above three, the World Federation of Small Landholders (La Via Campesina) demanded a revision in the World Bank's approach 'which operates on the principle of land for those who can afford it' (DSE 2001: 26). This organization and the Food Information and Action Network (FIAN) expressed the fear that 'the market-assisted approach is leading more to a reconcentration rather than redistribution of land'.

Privatization of customary land in Malawi, Sudan and Uganda

Despite existing strong arguments for maintaining customary land tenure, including those of the World Bank itself, and in spite of a lack of hard evidence on the production superiority of private/freehold/individual tenure over communally owned customary landownership, a privatization policy under neoliberalism reforms is pursued with vigour in some Latin American and many African countries. It is also enforced in spite of empirical evidence that the secure and inheritable customary tenure system is as compatible with the production of export crops as with food production. It is also compatible in production incentives and risk reduction (World Bank 1987, 1992: 142–3; Feder and Noronha 1987; Feder and Nishio 1997).

In a carefully conducted empirical analysis of a sampled survey of private (*malio*) land and customary landholdings in Uganda, Combya Assembajjwe *et al.* (2002) and Place and Otsuka (2001: 105–27) show that customary land tenure has, for a long time, enjoyed statutory protection and that there are no differences in investment incentives between the two systems in terms of coffee planting. Yet, the recently instituted Land Act of 1998 in Uganda stipulates the conversion of customary land into private (freehold) land property at a time when a large number of wealthy farmers obtained 'leased' customary land (Combya Assembajjwe *et al.* 2002: 154).

All available studies on land tenure systems in sub-Saharan Africa agree that land has for centuries been communally owned by indigenous groups (tribes, families and communities). The groups devised sets of working rules for land rights (use and occupancy) and subsisted by grazing and cultivating food crops when rain permitted. As recognized by anthropologists and geographers, this customary land tenure has been not only the most suitable socio-ecological system for land use and livestock husbandry in semi-arid agriculture, but also the cornerstone of food security and social security for millions of indigenous people. Likewise, its contribution to economic growth includes employment of the pastoralists' family members within a rational division of labour of women and children, the production of a considerable part of the countries' total meat, milk, wool and hides and the conservation of natural resources. Despite all that, privatization of communal customary land is given prominence in the current neoliberal land policy orientation.

Individualization effects

The findings of available studies on Malawi, Kenya, Uganda and Sudan suggest that individualization of customary land has led to:

1   The vulnerability of individual owners to the loss of land property by sale to larger landowners and urban land speculators, and by mortgage as a result of heavy indebtedness.

2 The weakening of women's customary rights in land and command over food.
3 The shift of resources away from food crops (mostly produced by small farmers) towards cash-exportable crops, e.g. tobacco. There is evidence from Malawi that land buyers converted the production of food crops into burley tobacco and that former small landholders became wage workers and net buyers of food. With the population growing fast, at about 3.1 per cent per year, food production per person fell rapidly during this period, while tobacco production increased from 70,000 tons to 110,000 tons and, by 1997, its area had increased substantially.[10] In his study of the rural credit market in Malawi, Mosley (1999: 377–9) found that financial liberalization led in 1994 to the abolition of the Smallholder Agricultural Credit Administration that targeted credit to poor farmers and subsidized small food growers (mostly maize). This policy has also resulted in the establishment of a World Bank-sponsored private entity (the Malawi Rural Finance Company), which finances tobacco growers instead of poor farmers and has thus disqualified food growers, who are among the poorest farmers. Mosley concludes that this credit liberalization policy 'has jeopardized the livelihoods of many poor people'.

Similarly, Adams (1997: 6) and Nsabagasani (1997: 33–6) show in their field studies in Kenya and Uganda, respectively, that a combination of privatization of customary land tenure arrangements and liberalization of the rural credit market has resulted in landlessness, increased rural poverty and loss of household food security. In the Masaka and Masindi districts of Uganda, nearly half the buyers of 108,500 hectares of privatized communal land were members of parliament, government officials and senior police officers. Once fenced, the ranches deprived pastoral households in the surrounding grazing areas of their use of traditional corridors for the passage of nomads and their animals in pursuit of the highly unpredictable rainfall. Similar problems were identified in Sudan by Abdalla (1993) in his field study of the agricultural land market in two provinces, Darfur and Kordofan. The falling food productivity associated with individualization of customary land tenure and credit liberalization that has not targeted the supply of credit to poor farmers is worrying indeed. The *FAO Production Yearbook* (1997) and FAO (2004b) show a post-1990 deterioration in food production per head and increasing numbers of undernourished people in sub-Saharan Africa. However, worsened food production and nutrition cannot be attributed solely to post-1990 credit liberalization and market-based land policy. They may also be ascribed to weather-related problems, etc.

Nevertheless, after these negative impacts of the sweeping post-1985 market-orientated policy, the World Bank has lately acknowledged, from accumulated empirical evidence on the impacts of individualization and formal titling of customary land tenure, that: (i) benefits from land titling are not enough to justify the costs of conducting land title investigation and land registry; and (ii) private titling of land is not a realistic alternative to communal tenure (Deininger and Binswanger in de Janvry *et al.* 2002: 417–18).

Market liberalization impacts on existing state-administered land reform

In this last section of examining neoliberal market orientation, we discuss changes introduced to complete and partial redistributive land reforms presented earlier in Chapters 5 and 6. The countries' experiences discussed below are: the market-orientated changes in the Chinese complete land reform; and the partial reforms in Egypt, Honduras, Nicaragua and the Philippines, some of which have already been studied in Chapter 6 prior to the adoption of structural adjustment and market liberalization. The section ends with a brief identification of the changes in Russia, the mother of communist Marxist agrarian structure that has influenced many developing and East European countries.

China

To remind ourselves of the major institutional changes after Mao's death in 1977 that laid the foundations for successive reforms, the changes are briefly outlined in three subsets of institutional arrangements. The first was the introduction of households' production responsibilities to replace the distribution of the commune/team collective earnings based on working points earned, out of which the individual beneficiary pays land taxation; the balance of crops produced is to be sold freely outside the state-controlled apparatus. The second is the expansion in private plots (as distinguished from collectives) for the production of household food needs. The third and most major change is the restriction of rural outmigration to urban areas and limiting the number of married people's children for the purpose of matching aggregate food demand to supply, and maintaining a low rate of agricultural labour force growth at 1.5 annually, while avoiding open unemployment or labour surplus.

Our discussion of China's complete land reform in Chapter 5 was concluded by raising the question: can the low-income economy of China, with the largest population in the world, sustain the achieved high agricultural output growth (7 per cent in 1978–82)? Can it also maintain the realized equitable distribution of productive assets and income in rural areas (land inequality Gini index was 0.21 in 1980)? And can the post-1978 economic reforms including certain elements of market orientation be utilized as complementary to central planning and not as an alternative? These issues are explored in the following discussion with an emphasis on whether and how the rural poor benefited from this transition during the period 1980–2000.

Considering that 70 per cent of China's total population is rural, the available information suggests that the complete land reform – introduced in 1948–50 at an early stage of national development – has been a leading support to the post-1980 realized economic growth and human development. The contribution of investment in human capital to output growth was estimated at 33 per cent in 1953–77, falling to 14 per cent in 1978–2000 (Perkins and Yusuf 1984; Chen and Wang 2001: Table 4). Researches have also agreed that, according to the results of the Rural Household Survey of 2000, inequality in the distribution of income/ consumption and between rural regions has deteriorated. Some have found that the worsening of inequality increased poverty and created a class of new rich (*wanyuanhu*) whose average per capita income in 1980–94 was 10 times that of the average rural population (Aiguo 1996: Tables 3.2 and 3.3). This emerging social stratification and inter-regional disparity are attributed partly to declining state investment in agriculture from 13 per cent of total budget expenditure in the 1970s to only 3 per cent on average in 1980–92. This rising inequality has led to large-scale rural migration to urban areas. Following the state-adopted policy on village–town expanded activities in industry and commerce, it was estimated in 1995 that 80 million poor peasants were seeking higher wages and better social services outside agriculture. The latter were effectively guaranteed by the agrarian communes before the introduction of some elements of market-orientated reform (Moldavian 1998, in Szelenyi 1998: 113 and Figure 5.5). The urban–rural income ratio has increased quickly since 1985 and, in a short period of 5 years, 1981–85, it exceeded the highest pre-reform level (Table 7.1).

Such a widening gap, combined with the collapse of people communes that have kept Chinese peasants on the land since the 1950s, has recently created serious social problems: crime, unemployment, severe housing congestion, child labour, prostitution and drug use. What about changes in rural poverty? Worsening income inequality from a Gini index of 0.21 in the 1970s to 0.29 in 1981 and further increased to 0.30 in 1995 and 0.34 in 1999 has adversely affected poverty prevalence in rural areas. Hence, the benefits of the remarkable record of economic growth at a high annual average of 9.7 per cent in 1978–99 are unevenly distributed among rural people in different regions. In fact, the worsening inequality contributed to a slight increase in the poverty level by 3.9 per cent in 1990–99. However, other estimates made using different poverty lines indicate that the headcount ratio in rural areas has not increased and has instead declined during this period at different rates (Chen and Wang 2001: Tables 3, 4 and A1).

We conclude that, because of its gigantic size, China's gradualist approach to agrarian structural reform is of international development interest. It has maintained socialism with its state-owned enterprises and planning mechanism, blending it with some elements of capitalism or market economy but *without* private land property. What has been allowed is land use rights in an average area of 0.6 hectares per household. In this gradual change, individual and household market incentives and risk-bearing have been eminent, and income inequality has been a high price to pay for sustained high growth associated with a high prevalence of social unrest. Despite the tangible material gains of most of the population, it was estimated, in 1996, that 1 million millionaires and their conspicuous lifestyle of capitalism are 'in sharp contrast to the 65 million destitute living in absolute poverty' (Aiguo 1996: 74). The nearly two to three decades of rural institutional transition from collectives and the commanding rural communes has given both the rural economy and the peasants the time to adjust, and the rural economy the chance to distribute the benefits from sustained farm and non-farm output growth

Table 7.1 Per capita income of rural and urban households, 1957–90

	Per capita income (yuan)		Ratio of urban to rural incomes	Rural income as a percentage of urban incomes
	Rural	Urban		
1957	73	254	3.48	29
1964	102	243	2.38	42
1978	134	316	2.36	42
1979	160	377	2.36	42
1980	191	439	2.30	44
1981	223	500	2.24	45
1982	270	535	1.98	50
1983	310	537	1.85	54
1984	355	660	1.86	54
1985	398	749	1.88	53
1986	424	910	2.15	47
1987	463	1,012	2.19	46
1988	545	1,192	2.19	46
1989	602	1,388	2.31	43
1990	630	1,523	2.42	41

Source: Aiguo (1996: Table 3.2).

to the majority of the rural people. All in all, I must say that this type of transition is a distinctively Chinese reform.

Egypt: deregulating tenants' protection and prevailing market-determined rent

Since the adoption of the World Bank- and IMF-induced price liberalization and structural adjustment programme in 1991, there has been a shift away from both the guaranteed payment of farm land rental values below market rates and the protection of tenants from unlawful eviction and towards market-determined lease arrangements. We recall from Chapter 6 that the reform programme of 1952 and 1961 has substantially reduced land rental values, in real terms, by fixing them at seven times the already low land tax that was assessed in 1947. The pro-market policy shift was introduced by Law No. 96 of 1992 on the relations between landowners and tenants, suddenly enacted in response to the lobbying strength of large landowners and supported by business people in industry and trade. This legislation abrogated the articles in the 1952 law protecting tenants from eviction and from raising the stipulated rental value. The 1992 law also increased the rental values from seven to 22 times the land tax until 1997, to be determined thereafter by market demand and supply forces. Accordingly, by 2001, 51,000 tenants had been evicted from their legally leased holdings, and only one-fifth of them had been allotted 2 feddans (2 acres) each from public land; the rest have become hired landless workers. Also, annual rent payment has greatly increased beyond the inflation rate from an annual average of E£500 in 1995–97 to E£3,500 per feddan in 2006. In a sense, this action by the legislators represents an urban bias because the fixed rents of urban housing have remained unchanged since the 1960s.

In the meantime, and despite sizeable rural–urban migration within Egypt and to the oil-rich Arab states, landless agricultural workers have increased from 4.2 million in 1951 to nearly 5 million in 1999. Alarmingly, there was a downward trend in agricultural real wages between 1986 and 1999: the average real wage in 1999 was merely 41 per cent of its 1986 level, resulting in higher inequality of household consumption distribution. Poverty prevalence in rural areas increased from 16.1 per cent in 1982 to 23.3 per cent in 1996 and was higher again in 1997 at 28 per cent (El-Ghonemy 2003; Datt *et al.* 2001).[11]

The substantial increase in rental values had worsened the inequality of household consumption distribution and landholding distribution. As shown in Table 7.2, after the market orientation policy that began in 1991, there was a considerable increase in land concentration from a Gini index of 0.43 in 1981 to 0.67 in 2000. Table 7.2 also shows that deteriorating inequality is manifested in the worsening shares of small farmers (less than 5 feddans or nearly 2 hectares subgroup) in the total area of landholdings, from 52.5 per cent in 1981 to 46.9 per cent in 2000. In the meantime, there have been rising shares of middle-income and rich farmers in the size groups of 10–50 feddans and 100 feddans and over, who benefited from output growth and crop price liberalization. Both changes have contributed to the above-mentioned rise in rural poverty levels. But, unlike China, Egypt's economic growth has slowed down sharply from an annual average of 8 per cent in 1982–83, falling to 4.6 per cent in 1990–2000 and further to 3.5 per cent in 2000–04 (World Bank 2002, 2006).

This trade-off between equity and agricultural output growth was examined by El-Ghonemy (1953, 1992: Table 2) during the two periods 1913–51 and 1952–86. During the first period, rental values were determined by market forces, and the British, French and Egyptian landlords had privileged access to credit and introduced technological change. These benefits are in addition to output gains from a

Table 7.2 Landholding distribution in Egypt, 1981–2000

Size in feddans	Shares in total area (%)		
	1981	1989	2000
< 5	52.5	48.9	46.9
10–50	18.1	19.0	19.1
100 and over	9.8	11.6	11.1
Gini index of land concentration	0.43	0.65	0.67

Source: The shares are calculated by the present author from the results of the Agricultural Census (*Nata'ig al-tidad al-Zera'i*) for the relevant years, Department of Agricultural Economics, Ministry of Agricultural, Cairo. The Gini index is calculated from the shares of number and area of all size groups from the same source.

Note
One feddan equals almost 1 acre (1.038) = 0.42 hectares.

substantial public investment in irrigation expansion, whereby the cultivation of cotton expanded and became the engine for export-led growth linked to the world market.[12] The upward trend in land profitability induced high rental values and land concentration (Gini index of inequality ranged between 0.62 and 0.77). During this period of market-determined prices (1913–51), the numerous tenants and landless wage workers were blocked by barriers to entry into the land purchase market. They had virtually zero opportunity to buy land, as we shall soon find out. During the second period, 1952–86, government intervention reached its peak, and there was an outright transfer of real income by 10–15 per cent from absentee landowners to tenants operating the land. In the meantime, agricultural output value lost its influence as a positive and significant determinant of rent according to economic theory (rent being the surplus over the cost of bringing land to cultivation), out of which tax is paid. What has happened to equity in relation to resource use efficiency and land output value since the 1992 deregulating reform and the role of land tax in these changes deserves further research.

The likelihood of agricultural workers purchasing land under market forces

I have examined elsewhere (El-Ghonemy 1993, 1999, 2005) the prospect for a landless worker purchasing agricultural land under market forces over a long period in Egypt and Syria. Despite their major differences in sources of irrigation, population pressure on land and the quality of arable land, both countries suffer from rapid shrinking of arable land per working person in agriculture. Furthermore, with this increasing scarcity of arable land, I have inquired into the likelihood of a wage-dependent agricultural worker purchasing a piece of 1 acre of agricultural land at the market rate in Egypt, as suggested by data given in Table 7.3. He or she would require to accumulate his or her entire wage for a period of 30 years on average, between 1977 and 1998, or 60 years if he or she saves half their wage earnings and spends half the wages on food and other necessary items, i.e. beyond the average life expectancy of 55 years. We should also note that neither a small tenant nor a hired worker in Egypt and Syria has access to institutional credit, which requires land property title as collateral in addition to the Islamic condemnation of mortgages. With regard to the likelihood of purchasing land through the market mechanism in Syria, I found from the results of available studies during my 2004 visit that the average daily wage in agriculture was S£150 and the average number of days worked was around 120 a year. Given the average total annual wage of S£18,000 and that the average price of 1 hectare of rainfed land was S£200,000, these facts suggest that the worker had to accumulate his or her wage earnings for an estimated 33 years to purchase 1 hectare of land in a rainfed area and nearly 83 years for an irrigated hectare (1 hectare equals 2.47 acres; and US$1 = S£11.5, official rate, and S£42 in the parallel market in 2001 (*Rapport Syrien Economique* 2001: B119).

Table 7.3 Average land values and daily wages in Egyptian agriculture, 1930–98

Values in Egyptian pounds and current prices	Market forces period		Land reform period	Liberalization policy and market forces period	
	1935–40	1945–51	1952–56	1975–80	1997–98
Market sale price (P)	119	415	180	3,000	35,000
Annual rental value (R)	7.1	22.7	17.5	29.5	1,300
Adult male wage (W)	0.029	0.102	0.110	0.365	8.000
Deflator (1966–67 = 100)	25	56	69	270	2,540
R as a percentage of P	6.0	5.4	9.7	1	3.7
P in year's rent	16.7	18.5	11.1	101.7	26.9
P in year W[a]	19.5	19.4	7.8	39.1	21.0

Sources: Rental values for the periods 1935–40 and 1945–51 are from El-Ghonemy (1953: Tables 10 and 37). The rest of the rental values are from the Egyptian Ministry of Agriculture's *Bulletin of Agricultural Economics*, several issues (in Arabic). Adult male wages for 1937–51 are average rates collected by the author from 98 villages in Lower Egypt and 83 in Upper Egypt during his work in the Fellah 'Rural Development' Department. The rest are from Radwan (1977: Table 3.2) and the Ministry of Agriculture. Wages during 1997–98 are from the Department of Agricultural Economics, Ein Shams University, Cairo. The deflator is from the CAPMAS *Statistical Yearbook*, Cairo.

Notes

One Egyptian pound equalled US$4.13 up to September 1949 and was devalued to US$2.87 until 1977. Between 1982 and 1986, it equalled US$1.22, and between 1997 and 1998 it equalled US$0.34. The sale price and rental value are for one feddan of land (0.42 hectares or nearly 1 acre).
a Assuming 210 working days per year. The deflator is the cost of living index for rural areas established by the Institute of National Planning, statistically linked with 1966/67 = 100.

Honduras

In my personal interview with Rigoberto Sandoval, who was the Director of the Agrarian Reform Institute in his own country, Honduras, before becoming my colleague at the FAO of the United Nations, Rome, he explained how much he had resisted the political pressure and attractive financial aid in the early 1980s to convert the agrarian reform programme from state-administered programme since 1973 into a market-based one, for his fear of potential landownership concentration. Yet, a land titling scheme was initiated by the USAID in the mid-1980s, supported by a credit supply provided by the World Bank and the Inter-American Development Bank. This action was followed in 1992 by the promulgation of the 'Modernization of the Agrarian Sector Law', prohibiting the expropriation of private landownership for its redistribution among the peasants and permitting the sale and leasing out of the agrarian reform beneficiaries' land.

Sandoval's fear was justified with regard to impending land concentration and small farmers' loss of their landownerships. Based on the results of the census of agriculture, land concentration after the redistributive land reform declined from 0.78 in 1974 to 0.66 in 1993 and, while we await the results of the 2004 census, several studies on the impacts of both the land titling and the 1992 agrarian modernization law indicate that, by 1997, nearly one-tenth of the government-mandated agrarian reform beneficiaries (since the 1970s) had sold their land at high prices. The buyers were mostly agricultural business people and such MNCs as the Standard Fruit Company (Carter 2002: 267–8, Table 10.2). On this relapse, Carter says, 'the tendency observed is that big landowners acquire the fertile land from small and medium farmers, but small farmers do *not* have any possibility of buying land from the large landowners' [. . .] 'Capital and other constraints limit the participation of small farmers (in the land market) nor can they effectively utilize the land market as a device to maintain or increase their access to land' (Carter 2002: 269; words in parentheses are mine).

Nicaragua

This was the second Central American country to adopt market-assisted land reform in the 1990s after implementing a state-led partial redistributive programme instituted by the Sandinista revolution in 1979. The beneficiary households of the 1979 agrarian reform represented 23 per cent of the total agricultural households, and the redistributed land was 28 per cent of total agricultural land area in the 1980s (see Chapter 5, Table 5.1). The post-1990 administration adopted a market-orientated land reform as an integral part of macro-economic liberalization, which followed the path of Honduras. The standard recipe of the market approach began with land titling to make land a collateral in the credit market. The new policy also included the privatization of state farms, the disintegration of collective cooperatives and the settlement of legal claims for land restitution. Owing to the long period of civil unrest since the Somoza administration (1936–79) followed by the Sandinista revolution, the post-1990 land policy gave priority for land allocation to the former state farmworkers, the 'ex-soldiers and the contra' who participated in the civil conflict and political disputes that ended the Sandista administration.

Like Honduras, available studies suggest limited transactions in the land sales market, owing to imperfections in the agricultural credit market, but greater opportunities for the peasants to access the land rental market, estimated at 22 per cent of total farm producers. From the results of a statistical analysis of the functioning of land markets in 1990 and in 1995, Deininger and Zegarra (2003: 1397) conclude that the land sales market was 'associated with land property transfers from small to large producers, that are likely to be economically inefficient'.

The Philippines

In Chapter 6, the partial set of land reforms implemented between 1954 and 1988 was briefly reviewed. In the face of increasing landlessness, inequality in land distribution and the strengthened power of landlords and the MNCs (or agribusiness), the concern over the slow progress in redistributive land reform and related social unrest continued into the 1990s. In March 1990, I was invited to participate in 'The Colloquium on Agrarian Reform' organized by the Institute of Agrarian Studies of the University of the Philippines at Los Banos and financed by the FAO.

Among the key empirical problems and policy issues discussed was the increasing opposition of landlords and MNCs to redistributive land reform programmes, the last of which was the 'Comprehensive Agrarian Reform Program (CARP)', initiated in 1988 by President Corazon Aquino, whom I met during the conference to clarify the objectives of CARP. Despite being enacted by the parliament, the landlords' and MNCs' lobby opposed the 'expropriation' of private land beyond the ceiling limit, which was raised from 7 hectares to 11 and 14 hectares. The target area for redistribution was reduced by 500,000 hectares (from 3.3 million hectares to 2.8 million hectares) They wanted the government to concentrate much more on the distribution of state-owned land and far less on the redistribution of private land property rights. Also, they demanded that the government acquire land for distribution through landowners' voluntary offer for sale at market value. A fundamental problem raised during the colloquium's inquiry was the almost lack of a complete system of landownership registry and land taxation. In addition, the meeting discussed the necessity of an effective participation of NGOs.[13]

Following the deliberations of the colloquium and the scholarly papers presented by eminent Filipinos, a detailed assessment of land reform implementation was conducted in 1996–99 by Borras, Jr and both the Department of Agrarian Reform and the Presidential Agrarian Reform Council. The assessment has led to the extension of CARP to the year 2004, and to its modification, accepting the World Bank model of negotiated land reform as a pilot project. The accepted market-orientated arrangements include: (i) the reclassification of land that is prone to redistribution from 'compulsory acquisition' to lands 'offered for sale to government' or 'voluntarily sold to beneficiaries' not through the government; (ii) accepting landlords' appeals to regain their expropriated land property through the judicial system; (iii) land reform beneficiaries to lease or contract production operations with MNCs and to follow their established arrangements; and (iv) the enforcement of both compensation payments to landlords and the assessed land value for sale.

Owing to its recent implementation, the results of the Philippines' market-based phase of land reform cannot be judged objectively. However, such experts on the country's land reform as Putzel are able to make useful remarks. Using available data from the Department of Agrarian Reform in 1999, Putzel (2002: Tables 2 and 3) found that only 7 per cent of the 1990 targeted total land area of compulsory acquisition for redistribution was realized during a period of 9 years. This slow pace of land property rights transfer was also observed in voluntary land transfer, whereby 33 per cent of the total targeted area of land below 50 hectares was actually transferred to beneficiaries. Putzel adds:

> In the Philippines, through various formulas to determine market value for compensation and for sale, the market-based programme has not only over-compensated landowners for generally poor lands, but also opened the way for fraudulent transfers [. . .] The World Bank's formula for negotiated land reform along with the government voluntary land transfers programme lack any attention to the political and institutional dimensions of the market place which are likely to undermine serious redistribution without significant state intervention.
>
> (Putzel 2002: 222)

Russia

I end the discussion on market liberalization related to land access with the post-1991 land privatization in Russia, the fountainhead of Marxist–socialist ideas of land policy, particularly the massive collectivization in Russia during 1929–35 and the spread of the communist system in several countries. The interest in the 'decollectivization and individualization' of land in Russia began in 1991 when I was invited by the former Soviet Union Academy of Agricultural Science (VASKHNIL) at Moscow, with funding from the European Union at Brussels. My task was to examine and advise on possible arrangements for land individualization and privatization. It is hoped that the inclusion of Russia in this chapter may help to understand the recent process of transition in several East European countries whose socialist systems were imposed on their rural people during the 'occupation' by the former Soviet Union.[14]

As the reader may know, the ruling years of former President Gorbachev, 1985–90, witnessed attempts to reduce bureaucracy in agricultural activities and provide farmers with production incentives to increase food supply via several attempts to reform property rights. Unlike China, the Russian post-1990 agrarian transformation started with the western concept of land property rights towards capitalism being central to President Yeltsin's policy about the restructuring of Russia's huge agrarian sector. Bureaucracy and centralization of decision-making were observed during my attempt to understand the procedure for identifying the original landowners of the collectivized farms, as the local officials concealed information on land property of former owners. They also resisted and opposed the new ideas on the individualization of collectives, the disbandment of state farms and the restitution of land property rights to the present generation of former owners. This agrarian transformation was stipulated by President Yeltsin's legislation of December 1991 on the Reorganization of the Kolkhozes and Sovhozes 'within one year into private enterprises with the right of individuals to purchase, sell or mortgage land without any restriction' (Nickolsky 1998: 198). I expressed to my hosts in VASKHNIL my disagreement and surprise about the feasibility of 1-year implementation period when it was suggested during my visit. Yet the implementation of this law went on, and it proved extremely difficult if not unworkable, owing partly to ambiguities arising from the lack of a precise mechanism and partly to the unwillingness of local officials to individualize giant collective farms in order not to lose their long-enjoyed financial and political privileges. Also, after such a long communist dominance, there was simply no experience with the land market. In addition, there was a lack of start-up operational capital in individualized farms, considering that such means of production as fertilizers, insecticides, seeds and farm machinery were manufactured and managed in giant state-owned enterprises that could not be separated operationally from collective landownership. Solving the problem of delivering the production requisites to individual farmers required further detailed regulations that could not be comprehended by ordinary farmers who, in practice, could not rapidly enjoy the benefits of spontaneous privatization decreed by President Yeltsin in 1991–93, which was supported by the IMF and the 'liberal' bureaucrats in contrast to the strong opposition of the 'communist' faction in the state structure.

We now have some idea of the transition experience. From his case studies in three districts in 1993–94, Nickolsky reports that peasants have been freed from central planning and tight government control, and that the communist agrarian institutions have been dismantled but not replaced by production assistance. He says, 'Farmers no longer know which government agencies they must deal with', they are uncertain about future property relations . . . 'The implementation was messy'. His empirical findings show that chemical fertilizers used in 1994 were only 11 per cent of the quantities used in 1986–90 (Nickolsky 1998: 202–6, Table 8.5). At national level, production of cereals in 2003 was 22 per cent lower than the 2000–02 average, and food production in 2002–04 was 10 per cent less than the average in 1992–94. Furthermore, the Gini index of land distribution inequality, which had been egalitarian for many years, was high at 0.46 in the year 2000 (FAO 2004a), and the average annual rate of total output grew at 6 per cent in 2000–04 compared with China's nearly 9 per cent, without privatizing land property rights.

Suggested reading

Carter, M.R. and Olinto, P. (1996) 'Does Land Titling Activate a Productivity-promoting Land Market?' Working Paper, Department of Agricultural Economics, University of Wisconsin, Madison, WI.

Deininger, K. (1999) 'Making Negotiated Land Reform Work, Initial Press. Experience from Colombia, Brazil and South Africa', *World Development*, 27(4), 651–72.

El-Ghonemy, M.R. (1999) *The Political Economy of Market-based Land Reform*, UNRISD Working Paper, Geneva: UNRISD.

de Janvry, A., Gordillo, G., Platteau, J.P. and Sadout, E. (eds) (2002) *Access to Land, Rural Poverty, and Public Action*, Oxford: Oxford University Press.

Ramachandran, V.K. and Swaminathan, M. (eds) (2002) *Agrarian Studies, Essays on Agrarian Relations in Less Developed Countries*, New Delhi: Tulika.

Szelenyi, I. (1998) *Privatizing the Land: Rural Political Economy in Post-communist Societies*, London: Routledge.

Wade, R. (1990) *Governing the Market: Economic Theory and the Role of Government in East Asian Industrialization*, Princeton, NJ: Princeton University Press.

8 Challenges and prospects

It is difficult for the author of a study such as ours to decide precisely what to say in the concluding chapter, after attempting to make each chapter comprehensive. Perhaps it is useful to begin with the common policy objective of poverty reduction, and to perceive the countries' past experiences presented in the case studies in Chapters 5, 6 and 7 as a set of responses to the big challenge of absolute poverty reduction in rural areas. Based on the empirical evidence accumulated over the past half-century, we then challenge the assumptions/suppositions used in the recent arguments for the policy shift away from state-administered or genuine land reform (SALR) and towards a post-1990 standard recipe of market-orientated land policy. This is followed by the challenges and dilemmas facing developing countries' governments, non-governmental organizations (NGOs), as well as international aid agencies and donor countries. The central question here is how a complementarity between a market mechanism and active public actions (governments and NGOs) realizes poverty reduction and social justice in rural areas.

This chapter concludes by addressing three questions: (i) if the present trends indicated in Chapter 1 continue, what would be the prospects for the hundreds of millions of undernourished rural poor?; (ii) from empirical evidence, how could an interdependence between SALR and market-based land reform (MLR) combined with rural non-farm employment expansion and partnership with aid agencies work in reducing poverty and land access inequality?; and (iii) in the current conceptual confusion and ambiguity about the dominant neoliberal/neoclassical economics policy choice of no government intervention where land concentration, poverty and malnutrition prevail in rural areas, what can social scientists contribute conceptually to enable the state and the market/capitalists to work together towards meeting the millennium development goal of halving poverty by 2015?

The pace of poverty reduction according to types of land reform

Our understanding of the initial agrarian conditions and the justifications made for the types of land reform discussed in the case studies suggest that rural poverty reduction is a common objective of both SALR and MLR. Given this key policy objective, is the scale of redistributing landholdings, including the beneficiaries' command over their food needs, the major determinant of poverty reduction? If not, what other factors have contributed to the reduction?

To probe these closely linked questions and subject to the availability of inter-temporal data, an intercountry comparison has been made with special emphasis on the absolute numbers of the rural poor, the size distribution of agricultural land and changes in food productivity, nutrition per person and agricultural workforce productivity. Considering the various limitations already alluded to in the case studies, this comparison is useful to policy-makers, development analysts and students of development economics/agricultural/rural development, because it indicates the probable order of magnitude of the time required to reduce poverty under different implementation conditions of land policies, and it draws some broad lessons learnt over almost half a century of experience. Moreover, I am aware of course that 'agricultural ' and 'rural' are not synonymous and that the data on the size distribution of agricultural land relating to the World Agricultural Census standard definition of holdings per household do not take into account land quality differences.[1]

With these clarifications in mind, I have elsewhere (El-Ghonemy 1990a: 167–78; El-Ghonemy *et al.* 1993: 359–62) conducted statistical analysis of the relationship between rural poverty, the inequality of landholdings distribution and agricultural output growth using data from a sample of 21 developing countries, including most of those in our case studies presented in the preceding chapters. The former was a simple regression analysis of the measurable consequences of land distribution or the Gini index of land concentration and agricultural output growth per head, being the independent variables influencing headcount rural poverty ratio over the 10-year period 1973–83. The purpose was to measure the extent to which variations in rural poverty and agricultural growth are dependent on the degree of land concentration. The latter study of 1993 was a refinement of the earlier 1990 analysis and had the purpose of estimating the probable order of magnitude of the time required to reduce rural poverty by: (i) relying on growth alone (without a land reform policy changing the degree of land concentration); and (ii) decreasing the inequality of land distribution (the Gini index) by way of redistributive land reform.

The results show that there was a positive but statistically insignificant rela-tionship between the level of agricultural GDP per head and land concentration and the negative sign between poverty relationship with growth (at the low elas-ticity of –0.27), meaning that, at the optimistic rate of 3 per cent annual growth per head of the agricultural workforce, it will take up to nearly 60 years to reduce the current rural poverty ratio by half.[2] The same poverty reduction could be at-tained much faster by redistributive land reform (SALR), which reduces land concentration by one-third, that is 14 years in the case of Egypt, by fitting their data in the equation, where the poverty headcount was reduced in rural areas after the 1952 and 1961 SALR from 56.1 per cent in 1951 to 24.4 per cent in 1965, instead of waiting 60 years until the year 2010. In broad terms, these results sup-port those obtained by Bardhan (1985: 90–2) and Saith (1989: 14–15, Table 16) using different sample data, but disagree with those of Griffin and Ghose (1979: 370–2), who believe that reliance on agricultural growth either has had no effect on rural poverty in Asia or, worse, may actually have accentuated poverty. Thus, differential decreases in land concentration by partial or complete land reforms could have a substantial and more immediate impact on reducing rural poverty than reliance on agricultural output growth effects.

The case studies presented earlier confirm these relationships. The *complete* land reform programmes in China and South Korea reduced poverty headcount by nearly 15 per cent per decade. Also, Iraq, whose land reform reduced the inequality index (the Gini coefficient) by two-thirds (from 0.90 in 1958 to 0.39 in 1982), achieved a faster reduction in rural poverty from about 70 per cent to 17 per cent during the same period, i.e. a 16.2 per cent reduction per decade. These substantial, rapid reductions in rural poverty were associated with a large-scale redistribution of agricultural land (over 60 per cent of total agricultural land area) compared with all India's redistributed area of 3 per cent of total agricultural land and a 4.7 per cent reduction in poverty prevalence in one decade after the implementation of its *partial* land reforms during the period 1953–79. The pace of rural poverty reduction in Pakistan, which redistributed 4 per cent of total arable land, was much slower at 1.6 per cent per decade during the period 1963–80.

Interestingly, the pace of poverty reduction in rural areas varies among states within the same country. In all India, the reduction was 4.7 per cent per decade, whereas in the state of Kerala, it was faster at 6 per cent, and the redistributed land was 17.5 per cent of total agricultural land area compared with the 3 per cent in all India. In addition, Kerala had greater investment in human capital, particularly education, combined with faster non-agricultural employment growth in rural areas and higher wage rates induced by strongly unionized labour. In both cases, the absolute numbers of the rural poor increased, owing to demographic and other variables operating in the economy other than land reform, as will soon be explored.[3]

Turning now to the corresponding effects of the MLR (market-based/assisted land reform), one finds it difficult to be assertive for two reasons. The first is the programme's recent occurrence in a sense of being *nouveau arrivé* and, from a long development perspective, its current implementation appears to be in an experimental stage. The second is the multiplicity of its interlinked constituents: repairing the credit market inadequacy for small farmers; relying heavily on external financial aid; arranging for land titling and property registration to make land collateral; improving the beneficiaries' production abilities and their farm management capacity in schematic pilot projects; involving local communities as mediators between voluntary sellers and willing buyers, etc.

As a close observer of this standard recipe since my earlier examination of market-based land reform (El-Ghonemy 1999, 2002), I have found that its poverty and distributional consequences are determined by effectively carrying out those constituents and by their sequencing in a dynamic rural economy, in addition to the response of the government machinery. However, the available scattered pieces of information presented in Chapter 7 indicate the slowness of progress made combined with a very limited scale of agrarian change and high transaction costs that are simply unaffordable by the landless wage-dependent worker and small tenants who are the intended beneficiaries.

Take, for example, the available information on the implementation of this approach around 1994–95 in both Brazil and South Africa. Despite the heavy financial support given (Chapter 7), the reported actual land purchasers (not necessarily the rural poor) by the year 2000 in Brazil were nearly 10,000 households out of an estimated target of 2.5 million, meaning an approximate 0.04 per cent of the total intended beneficiary households during the planned 5-year programme. Contrast this pace with state-led reforms in Egypt, Iraq and South Korea, where all of the intended beneficiaries received land during the same period. With regard to South Africa, empirical evidence shows that, after 3 years of carrying out the programme, supported by exceptionally high foreign aid and at the cost of an increased debt trap, only 20,000 beneficiaries purchased 200,000 hectares at a highly subsidised price against a planned target of 1.8 million beneficiary households or an intended average of 600,000 per year, meaning that less than 1 per cent of intended beneficiaries have purchased land. Probably the accumulated evidence from Brazil and South Africa, as well as from other countries, is behind the recent candid recognition by the designer and promoter of this land market approach that, 'the poor will often be unable to access land through the land-purchase market' (Deininger 2003: 150).

Non-land reform factors contributing to poverty reduction

Empirical evidence on country experiences presented in Chapters 5, 6 and 7 indicates that both types of state-administered and market-based land reform have been operating within dynamic forces in the national economy integrated in the rapidly globalized world economy. The effects of land reform on poverty reduction cannot, therefore, be isolated from their linkages with these dynamic forces. Apart from climatic fluctuations, important factors include population growth, rural outmigration, irrigation expansion, non-farm employment and remittance receipts. They are examined briefly in this section.

Population growth

Although it is widely recognized that the rate of population growth has a substantial effect on per capita income and the demand for food consumption, it is difficult to ascertain whether the combination of land reform and fertility reduction in such countries as China, South Korea, India and Tunisia was intentionally conceived, or whether population policy was a separate instrument required for overall national development. From what we know of the Chinese sequential events, socialist land reforms and the demand for food and strict control of population growth are linked. If we accept the pre-land reform poverty estimates used in Chapter 5, the intertemporal effect of combining complete land reform and population control is clear with regard to China, South Korea and Tunisia.

The dynamic forces in both the agricultural sector and the national economy, including demographic changes, have contributed to the fast pace of rural poverty reduction in China and South Korea, while Tunisia has the lowest level of both poverty and population growth in North Africa. The large and complete scope of Tunisia's land reform, benefiting nearly half the total agricultural households in the 1960s and redistributing property rights in 57 per cent of the total arable land, has been combined, since 1963, with an official but seriously implemented programme of family planning fostered by the late President Habib Bourguiba himself during the 1960s and 1970s. In this Muslim country, the manufacture and sale of contraceptives are heavily subsidized by the government, abortion and sterilization are legalized, and polygamy has been abolished. Taking 1960 as a reference point in time, life expectancy at birth increased sharply from 48 to 70 years by the late 1990s, food production per capita was 30 per cent higher, and a similar increase in agricultural labour productivity was achieved. These factors are likely to have affected the extent of rural poverty reduction (see El-Ghonemy 1993a: Tables 3.1 and 8.2 for the 1980s and 1990s; and FAO 2004b for the period 2000–02).

Technological change in agriculture

Given their continuing importance in both the national and the rural economy, the combined effects of land reform and labour-intensive technical change are important determinants of rural well-being. Remarkably, between the 1960s and the 1990s, irrigated land area was more than doubled in both the country with the largest population, China, and the country with the smallest population, Tunisia, associated with their land reforms. The substantial expansion in irrigated land has increased land use intensity and expanded opportunities for land access. Combined with high-yielding crop varieties, this technological advance has also been realized in such partial land reform countries as Egypt and Pakistan, which have reduced rural poverty far beyond their modest scale of land redistribution noted earlier in Table 5.1 (10 per cent and 4 per cent respectively). Following the construction of the Aswan High Dam, Egypt has, since the late 1960s, accelerated the reclamation and distribution of state-owned land among poor peasants and increased the ratio of land use intensity from 1.7 in 1960 to nearly 2 in 1990.[4] Pakistan, on the other hand, reduced land concentration from the Gini index of 0.54 in 1963 to only 0.52 in 1980, but expanded irrigated land area from 60 per cent of the total arable area in 1965 to 77 per cent in 1985, combined with yield-increasing technology during the so-called Green Revolution (1965–85). Irrigated land increased further to 84 per cent of total agricultural land area in 1999–2001. Furthermore, Pakistan managed to sustain agricultural growth at an annual average of 3.5 per cent annually.

Clearly, the gains from technological advance in agriculture are not confined to the beneficiaries of land reform and other small farmers. Productivity gains and payment of subsidies for production inputs are mostly accrued to medium-sized and large landholders in Egypt and Pakistan, as in other countries with partial land reform, where the non-reform agrarian sector is considerable. For example, in El-Ghonemy (1993b: 67–9), I examined who benefits from these productivity increases in Morocco, whose land reform sector comprises only 4 per cent of total agricultural land area and 2 per cent of the total agricultural households even after instituting three land reform legislations in 1956, 1963 and 1973. I found that, while farm units in the large size category of 50 hectares and over represented only 0.7 per cent of the total number, their share was 17 per cent of the total land area. Yet, these fortunate large farmers have captured a disproportionately high share (43 per cent) of total government expenditure on the subsidization of production inputs (seeds, fertilizers, irrigation water charges, combine harvesters and labour-saving tractors that were subsidized at 85 per cent of their market value).[5]

Rural non-farm employment, migrant workers' remittance transfers and cultural factors

To illustrate the working of these factors, we should be country specific. Egypt and Pakistan are two Muslim countries which have undergone partial land reform. Both have been able to reduce rural poverty disproportionately more than their land reform's redistributive scale noted earlier. Significantly, the expanded international labour market in the oil-rich Arab countries is more likely also to have contributed to rural poverty alleviation by way of Egyptian and Pakistani agricultural landless workers' migration, seeking higher earnings at high transaction costs. This expanded demand in the Middle East labour markets has brought about a notable flow of remittances into the rural areas of the workers' home countries that has helped boost the incomes of the risk-taking landless poor; many of them have used remittances for housing construction and the purchase of a small piece of farmland. In addition, the expansion in non-farm employment opportunities *within* their own rural areas has diversified the source of their income and helped to reduce inequality of household income/expenditure distribution.

From their analysis of the results of Egypt's household income/expenditure surveys for 1997 and 2000, El-Leithy *et al.* (2003: Table 4.1) have found that wage-dependent agricultural workers are the largest category of employees, whose non-farm wage earnings represent 41 per cent of their total income, compared with 31 per cent from agricultural activities. The inverse and significant relationship between the amount of Egyptian migrant workers' remittances received and the poverty level (headcount index) was manifested in their wide fluctuations in the 1990s in connection with the 1991 Gulf War induced by the Iraq invasion of Kuwait. Total remittances, in real terms, peaked in 1990 (US$6.5 billion), dropped sharply to US$4.5 billion in 1991 and have continued their downward trend (owing to an active and competitive labour market which tends to replace many Egyptians with Sri Lankan and Pakistani workers), reaching US$2.5 billion in 2000. The rural poverty level has followed the fluctuating remittance movement; poverty increased to 28.6 per cent of the total rural population in 1991 and fell to 23.3 per cent in 1996 after the Gulf War ended (Adams 2003; El-Ghonemy

2003). Of course, in the context of macroeconomic dynamics, remittances cannot be held solely responsible for this sharp fluctuation in the rural poverty level.

Another significant determinant of the non-land reform changes in rural poor income in both Egypt and Pakistan is the annual payment of *Ushr* and *Zakat*, according to the household earning capacity mandated by Islamic principles. This form of just, progressive tax of net income (2.5 per cent) in favour of the disabled and poor (*Mustahaqeen*) is practised informally and individually in Egypt, but is institutionalized in Pakistan, whereby the payments are pooled by government machinery. It has been estimated that these payments regularly reach 45 per cent of the total rural poor households (Ali 1985).

To sum up, these non-land reform factors (or variables other than landholding distribution) are the residual in measuring the degree of correlation known in statistical analysis as the regression coefficient of poverty on land concentration and agricultural output growth. The results of analysis of the data collected from 21 countries cited earlier show, in broad terms, that the variation in poverty levels among the sample of countries is explained mostly and positively by land concentration (Gini index). An estimated residual at 30 per cent is considered to be the 'unexplained' portion of other variables operating in the economy that have been outlined briefly in this section.[6]

Suppositions challenged

Distorted perceptions based both on partial understanding of the complex determinants of rural poverty and on biased or unrealistic suppositions produce public actions that bypass the real determinants. They may also keep many of the present millions of rural poor in persistent deprivation and malnutrition, contrary to the countries' moral and political commitment to meet the millennium development goal of halving the average 1990–91 number of poor by 2015, considering that rural population growth is greatest in poor (low-income) countries where rural poverty is concentrated. In Chapter 7, I have challenged the assumed marketability of agricultural land, and I shall not discuss it further.

The first supposition or claim is that the obsession of both the policy-makers of developing countries and several development economists with state-mandated redistributive land reform (SALR) has led to inattention about agrarian benefits to be induced by activating the land market combined with land registration and formal titling. It is argued that these market-based institutional arrangements provide land with collaterality in the credit market as requisites for increased land access for the poor rural households. Accumulated empirical evidence presented in Chapter 7 indicates the slowness, high cost per beneficiary, including external financing, and minimal effect of the market approach as an alternative to SALR for poverty reduction. In addition, we have seen from evidence presented in Chapter 4 the advantages of small family farms over large farms, and that the redistribution from the latter to the former would most likely raise efficiency in resource use and enhance rural people's welfare. We have also seen from Chapter 7 that activating the land purchase market depends heavily on foreign lending for a sustained supply of credit that tends to increase the already heavy burden of foreign debts. Poverty reduction is conditional upon these financing reforms. Do poor tenants and wage-dependent landless workers have the luxury of waiting for the land and credit markets' imperfections and issues of political reliance on external financing to be resolved?

The second supposition to be challenged is the generalized insistence on access to land per se to alleviate rural poverty. In reality, it is not so much the land access policy as the scale and pace of redistributing landed property and the provision of complementary production inputs that determine the extent and speed of alleviating wealth inequality and rural poverty and, in turn, the rate of human capital accumulation. Related to this confused assertion is the claim of the re-emerged importance of the land lease market to fill up a missed opportunity in past redistributive land reforms. Empirical evidence suggests that most, if not all, these reforms have constituted the realization of tenure security for the millions of tenants, and these reforms have, since the 1950s, provided them with efficiency gains from tenancy regulations (e.g. Egypt, India, Pakistan, the Philippines, South Korea and Syria). These gains were a consequence of the land reform's removal of the landlords' institutional monopoly, which was replaced by state-guaranteed lease security and protection of tenants from unlawful eviction, which were enforced until the adoption of market liberalization policies in the early 1990s, when these guarantees were abolished. Moreover, the SALR programmes reviewed in our case studies have – since their inception – provided, at low transaction costs, the registration of rental contracts and necessary working capital from the village/district-level cooperatives, strengthened by technical support from government specialists (extension services).

This brings me to the third challenged supposition. It is that centuries-old customary land tenure and temporarily pooled resources for cooperative land use are not compatible with enhancing efficiency and equity requirements in the modern era of neoliberalism and economic globalization. We have seen in Chapter 7 and from the findings of carefully conducted field studies in sub-Saharan Africa (e.g. Platteau 1995, 1998; Mosley 1999; El-Ghonemy 1999; Assembajjwe *et al.* 2002; Sjaasted 2003) that the insistence on the enforcement of individualization and privatization is based on an oversimplified western standard prescription for a complex agrarian system, and is unwarranted because customary working rules (e.g. inheritance and the allocation of land use rights among men and women by village heads and tribal chiefs) have proved that this tenure system is efficient and as compatible with export crops as with food production. Research findings cited earlier have also shown that there is no difference in efficiency and investment incentives between the customary tenure system and the imposed privatization and collaterality of land in the credit market. Moreover, under the latter, small landholders have lost both their centuries-long food security arrangements and their privatized land title to non-agriculturalists and urban land speculators, and rural women have also lost their traditional land rights and management of household food security. Empirical evidence indicates that reconcentration of land holdings has emerged after obsessive individualization and land titling policies.

The fourth challenged proposition views genuine land reform (SALR) not as a dynamic policy, but as a static, once-and-for-all redistribution of the existing stock of wealth in land property and other material capital, at one point in time. Our study of the actual design and implementation of these reforms over the past half-century suggests that no single land reform policy is appropriate to all conditions. Nor does any policy, once chosen, remain static in the face of dynamic rural development and structural changes in the national economy. The Chinese land reform, instituted in 1948, has gone through several adjustments from equal individual landholding (0.15 hectares) during the period 1948–52, to collectivized cooperative landholdings and communes, to the household responsibility system, providing greater production incentives and rewards in 1978, leading to current gradual institutional changes, including the introduction of market-orientated elements, replacing the communes. Additionally, 10–12 years after their initial reforms, both Iraq and Egypt amended latent defects in their legislation by substantially lowering the prescribed ceilings. Between 1975 and 1980, Iraq realistically faced its problems in the institutional organization of production by liquidating many inefficient state farms and curtailing collective farms for lack of farmers' interest. We have also seen how the Philippines Presidential Agrarian Council in 1996 extended the comprehensive programme, initiated 15 years earlier, to the year 2004, and how the South Korean authorities allowed the land reform beneficiaries to rent out part or all of their lands in response to a changing rural economy (many of them became older in age and family labour was induced to migrate to urban areas and abroad, with an increasing inflow of remittances sent to the remaining household members).

Fifth, it has been suggested in the literature that the origin of SALR is incompatible with parliamentary and democratic rules of law. The cases of India, Sri Lanka, Chile (1967) and, most recently, Bolivia (2006) challenge this view. The Philippines and Brazil are currently implementing their land reforms under parliamentary majority. It is equally incorrect to claim that all authoritarian regimes bring about land reforms. Many do not. In some cases, authoritarian regimes have allied with big landowners, MNCs and industrialists, for example Paraguay, Bangladesh, Pakistan (after 1977), Nigeria and Chile (after 1973).

Sixth, our case studies of state-administered and market-based land reforms have shown that governments (and their technocrats) are not neutral as usually assumed in neoclassical economic theory and neoliberal models. Nor are they to be understood as Plato's puritan 'Guardians' endowed with high moral standards and meticulous ethics. Bureaucracy-related high transaction costs, exploitative relations and corruption exist under both types of land reform. Often in developing countries, the interests are served of those economic classes on whom governments depend for their tenure in office, while the interests of the rural poor, having no influential NGOs of their own, are passed over.

Lastly, but importantly, the evaluation of land reform policy should not be according to the evaluator's own set of criteria or what he/she thinks the objectives should be. Instead, it should be against the precisely stated objectives at the time of the reform enactment. The resulting confusion arises not only from his or her own ideology and personal beliefs, but also from following a narrow perception guided by the evaluator's partial understanding and specific professional field, using textbook jargon. Consequently, throughout my half-century-long academic and empirical experiences, I have learnt how a single country's land reform may be contradictorily judged by different evaluators with different ideologies.

In broad terms, social justice, expressed in different and simple parlance (Chapter 2), is the common objective, according to my study of countries' land reform experiences. For example, liberation of the exploited peasants and semi-feudal bonded labour was particularly emphatic in the objectives of the Mexican, Cuban, Bolivian, Indian and Peruvian reforms. Egypt, Iraq and Syria used the stock phrases of the 'abolition of feudalism and humiliation'. Several country leaders presented land reform as the impetus 'to raise the heads of the poor cultivators and to regain their dignity' (for example, Mexico's Zapata in 1911, Iraq's Qassim in 1958, Egypt's Nasser in 1952, Pakistan's Bhutto in 1972 and Bolivia's President Morales in 2006). In all these cases, as well as in many others, land reform was conceived as an effective approach to fairness, ending the harmful manipulation of land, credit, labour and grain markets to make high profits out of the miseries of small farmers, poor peasants and pastoralists which had arisen under institutional monopoly powers. It seems that policy-makers equated the economic meaning of the abolition or weakening of landlords' monopolistic power in the agrarian market structure with the ethical meaning of providing peasants with liberty, property rights, self-esteem and eventual command over their own food needs. Because no monopolist gives up power voluntarily, the break-up of concentrated landed property and the redistribution of property rights require state intervention. In such circumstances, the institution of government has to extend its limits of ordinary functions, whether by democratic or authoritarian means, to adjust property rights and limit the economic freedom of entrepreneurs for public welfare and the interest of the disadvantaged section of the rural population. In a sense, land reform combines the economic, moral and ethical meaning of justice.

If my interpretation of social justice as the common objective is acceptable, land reform is basically a corrective policy for past injustice. Since the work of John Locke and David Hume in the early seventeenth and eighteenth centuries, these paradigms of justice and welfare have been a fertile ground for all branches of social sciences. Recently, and from different perspectives related to property rights and poverty, they have constituted the work of Rawls (1972), Roemer (1996) and Sen (1981, 1982, 1997).

In Chapter 2, I discussed these philosophical ideas of social justice in terms of the enhancement of secure access to land for the rural poor to realize their productive potentials and to command their family labour and food needs. Thus, the emphasis placed by land reformers upon attaining social gains and political stability is well founded where it enhances the productive capacity and nutritional status of the beneficiaries, and removes institutional barriers to peasants' freedom and the well-being of the rural poor. If international aid and social sciences, including neoclassical economics and neoliberalism, are not about these inequality and poverty issues that millions of people talk about and fight for, they are in my judgement about nothing practically useful to social peace and human welfare.

Prospects and challenges

Chapter 1 describes how the recent trends that are unfavourable to the present generation of rural poor's opportunities for secure land access constitute a real challenge for rapid poverty reduction. The case studies presented in Chapters 5, 6 and 7 indicate that rising demand-driven access to land of poor landless and near-landless households remains a central policy issue in the battle against injustice, malnutrition and poverty in most LDCs' rural areas. The diminishing accessible opportunities to meet this rising demand are manifested in the increasing land concentration and continuing rural unrest and violent conflicts over land access. These conflicts are between the poor landless, invading uncultivated arable lands supported by NGOs and rural development scholars, on the one hand, and absentee landlords protected by their own guards and public police force, on the other (examples are Brazil, Colombia and the Philippines, to mention a few). Other types of land conflicts have resulted from the rush of governments, pressurized by international aid agencies and several donor countries, to privatize collective farms and communally held, centuries-old indigenous lands. The significant and decisive response to this challenge by the Bolivian leadership in early 2006 is fresh evidence that redistributive land reform never dies. In a sense, it is in defiance of the neoliberalist idea that government intervention in land tenure arrangements is irrelevant to promoting incentives for production enhancement and poverty reduction.

Before we explore accessible opportunities for the rural poor households' secure landholding (landownership purchase, lease, sharecropping and inheritance arrangements) and for the convenience of the reader, let us outline briefly the recent interlinked trends suggested in Chapters 1 and 3, the response to which represents the big challenge ahead.

The first and serious tendency that I have carefully documented is the increasing number of rural poor and malnourished in LDCs, particularly the landless and near-landless households (Chapter 1).

Second is the increasing inequality of landholding distribution, implying an increase in the monopoly/monopsony powers of large landowners in the land sale, land lease and labour markets, as well as in the use of scarce water and irrigation pumps in arid and semi-arid areas. As conceptually argued in Chapter 2 and empirically demonstrated in Chapters 3 and 4, these forms of institutional monopoly are associated with market imperfections, resource use inefficiency and skewed distribution of income/consumption. This continued high inequality has slowed down agricultural output growth (Chapters 1 and 3) at a time when poverty reduction requires sustained productivity growth. See the compelling evidence on the overriding importance of redistribution for poverty reduction in note 12 of Chapter 2 and note 2 in this chapter.

Third, the aggregate supply of cultivable land has declined, owing primarily to economic reform-mandated budgetary cuts to reduce fiscal deficits, including public expenditure on irrigation, drainage and soil conservation works – these were mistakenly considered as public consumption and not as necessary investment. This downward trend has, since 1985, been accompanied by a sharp fall in international aid (by most donors) from the 1996 commitment of 0.7 per cent of GDP to 0.2 per cent in 2006, including aid to agriculture/rural development in poor LDCs, implying a reduction in employment opportunities for output growth per unit of the rapidly increasing agricultural workforce. Another possible reason for the budgetary cuts–shrinking land supply trend is the speedy urbanization rate, resulting in the transformation of fertile land to non-agricultural purposes and the rapid rise in land prices exceeding the capitalized value productivity in agriculture (Chapters 1 and 3).

Fourth, in addition to the dynamic forces indicated above, shrinking employment opportunities within agriculture have been induced by the tendency towards less labour-intensive and more capital-intensive agriculture. This tendency, being unfavourable to the rural poor, has been accelerated by the widespread use of labour-displacing technology, facilitated by rising globalization, particularly the removal/relaxation of farm machinery import duties. These documented tendencies have been compounded by rising unemployment in urban areas in conjunction with the competitive replacement of unskilled rural workers by more skilled and educated job-seekers who are competing for low-earning jobs, leading to a likely fall in real wages.

Fifth, the capacity and authority of LDCs' governments to interfere in the capitalist agrarian economy has been largely curtailed by forceful western support for the land credit market path of voluntary landownership transfers, considered by its supporters as an easy political alternative to state-mandated redistributive reform (Chapters 2 and 7).

Dilemmas

Policy-makers of governments adherent to economic reform-linked conditionalities were faced with a dilemma. If they did not break up land concentration and its associated institutional monopoly powers (Chapters 2 and 3) on the scale required to rapidly reduce rural poverty in 2015 to half its level in 1990–91 according to their international commitment towards the first millennium development goal, they would be likely to face social unrest and continuing violent occupation of absentee landlords' uncultivated lands (as in Brazil, for example). They may also not be able to enforce long-term lease contracts at fair rental values demanded by tenants and protect them from unlawful eviction.

In this alternative, governments that are reluctant to intervene are more likely to receive exuberant appreciation by western donors and international aid agencies and some affiliated social scientists who claim to be value free. In the writings and prescriptions of anti-intervention sponsors, it is noticeable to see such phrases as 'genuine redistributive land reform is not politically feasible', 'not compatible with political reality', etc. One wonders by what authority these pro-neoliberal economists and other critical social scientists impose these value judgements on sovereign governments of LDCs.

Prospects

These value judgements and discernible bias are puzzling indeed. In this time of confusion and misinterpretation of the roles of, and boundary between, the market and the state in rural poverty and injustice reduction, one expects the social scientist to rise to the challenge of abating this confused reluctance over land reform. Whether land reform is politically feasible under certain power structures is not for the economist or the development analyst to judge. What he or she can do is to use the faculties and professional tools to understand rural underdevelopment problems in a specific situation, to analyse the land tenure structure and functions and define the determinants of poverty, chronic malnutrition or hunger and gross inequality of land distribution, and to suggest non-utopian alternatives, including selective state intervention for country-specific situations and not a universal prescription. Throughout this book, our study has not considered the docile maintenance of the status quo of agrarian injustice to be an alternative. Instead, this book argues that land reform is the alternative to a state of rural underdevelopment characterized by skewed distribution of land, income and accessible opportunities, as well as increasing poverty and undernourishment.

Keeping this in mind, we begin with a clarification of a common misinterpretation of 'for' or 'against' government intervention and 'pro' or 'anti' market approach *without* specification of the poverty reduction focus and the nature of government intervention. It is equally unhelpful to make a generalized policy preference between 'market-based' and 'state-based' (or administered/mandated) land reform *without* placing the debate in a country-specific agrarian context. For example, South Korea's initial agrarian conditions made it necessary for the country leadership to pursue, since the 1950s, a complete land reform policy combined with rural non-farm employment expansion and substantial human capital investment plus institutional arrangements to provide information and support and guide the market towards greater exports and investments in priority areas. In addition, the government has guided the market towards economic stability by such measures as setting an effective exchange rate and interest rate and promoting exports. Like South Korea, Japan and Taiwan of the Republic of China carried out, in the late 1940s and 1950s, extensive redistributive land reforms combined with the introduction of government regulations that enabled the market mechanism to gradually function effectively.

One should ask: in the absence of government intervention in these countries, what would have happened to rural poverty and inequality reduction and to the sustained economic growth, bringing about a superior economic performance, described by the World Bank (1993) as the East Asian Miracle? This complementarity between government activist and market-assisted policy at an early stage of development is behind these countries' current superior economic performance.[7] Likewise, China, after achieving an egalitarian rural development based on a complete land reform implemented during the period 1948–77, found it necessary in the 1980s to gradually enable market mechanisms to promote household production incentives and farmers' earnings. Hence, without a complementarity between an active market and government role in reducing land concentration and inequality of accessible opportunities, neoliberalism for free trade, higher growth and poverty reduction are basically ineffective.

Mobilizing the demand for this prospect's achievement and for monitoring progress made requires popular participation. In recent history and in many LDCs, motivated NGOs have expressed the landless and poor peasants' frustration with national and international obstructions to government's action for a speedy reduction in land concentration and rural poverty over the last two decades. They have also been critical of the capability and potentiality of the land market approach *alone* effectively to reduce landlessness and rural poverty (Chapters 5 and 7). Elsewhere (El-Ghonemy 1984b, 1986: Chapter 9), I have argued for popular participation as an anti-poverty strategy, and presented empirical evidence on how NGO pressure groups could help to mitigate land tenure injustice and violent unrest in rural areas through secure access to land (see also Ghimire and Moore 2001).

Through legal means, NGOs, particularly agricultural labour trade unions and peasant organizations, can exert pressure to restore land grabbed by influential persons or seized by moneylenders. In alliance with the committed middle-class intelligentsia, they can help to mobilize public opinion and gain support from the media locally, nationally and internationally. They can also collectively lobby for greater access to landownership and for reduction in exorbitant rental values and their enforcement.[8] Still, organized participation faces enormous barriers. Many governments either prohibit the very existence of agricultural trade unions, while allowing them in non-agricultural sectors, or render them powerless through denying their rights.[9] See Yunus (1994) on Bangladesh's pioneering work.

The effectiveness of non-statutory organizations is enhanced by occupational or gender homogeneity among their members. For example, women, who constitute almost half of the agricultural workers in many LDCs, have traditionally been excluded from village organizations, trade unions and agricultural cooperatives, whose membership has been the prerogative of men. With the exception of socialist countries, rural women are rarely given title to land in land reform and settlement schemes (Palmer 1985). This hardworking silent majority have been, and still are, denied their legitimate rights in land to influence the design of policies and programmes in order to reflect the realities of their high rate of participation in the total labour force.

A good example of potential NGOs' drive to provide land access to the landless poor is the Brazilian Rural Landless Workers' Movement (MST). Over the last 20 years, it has led the campaign for redistributive land reform through the mobilization of the landless agricultural workers to occupy and then receive property title for uncultivated arable land. Its leaders rely on a legal foundation provided by an article in the 1964 Constitution, declaring the uncultivable arable land to be eligible for redistribution because of the landowners' failure to assume their 'social responsibility'.[10]

The potential pressure of NGOs teamed with a committed intelligentsia represents a cheerful prospect in a gloomy situation of persistent rural poverty and governments' reluctance to act. The millions demanding accessible and equal opportunities for their well-being and demanding their legitimate rights in land know what they really want and should be heard. Among the sharecroppers, hired landless workers and nomadic pastoralists, men and women have perceptions that may differ from those of politicians and government programmers. In their own interest, politicians, programmers and foreign aid-giving agencies should listen to, and learn from, the rural poor through their own organizations. Excluded from effective participation for too long, once motivated, the poor can collectively articulate their perceptions of priorities and programmes. Generally speaking, the rural poor know neither how to lobby nor how the political system of their countries is organized.

The constituencies of rural well-being bring us to where the book began, Chapters 1 and 2, making a distinction between rural betterment and rural development. The latter embraces land reform integrated with popular participation, active government investment in education and health care, and expansion in both irrigation and non-farm employment activities in rural areas which aim not only to raise physical productivity but, importantly, to expand human capabilities (literacy, nutrition, life expectancy and dignified participation in village organizations for developing local communities). In this broader view of secure access to land for the rural poor, the emphasis is not on the narrow issue of freeing the market or for government intervention, but on the broad issue of freeing the landless, poor peasants and pastoralists from socio-economic deprivation created by the growing institutional monopoly powers, which are the ugly face of capitalism in the agrarian societies of developing countries. Leaders of LDCs with capitalist agriculture should look to land reform integrated in an anti-poverty rural development strategy as a means of solving the existing fundamental contradiction of excess accumulation of rural wealth with excess poverty.

Appendix A

Estimates of rural poverty prevalence in 64 developing countries

Country	Year of estimate	Total population (millions)	Proportion of rural population to total (%)	Percentage of rural population in absolute poverty	Estimated number of rural poor (millions)
Africa					
Benin	1979	3.4	70.7	65	1.6
Botswana	1982	1.0	75.0	55	0.4
Burundi	1978	4.0	97.7	85	1.6
Cameroon	1978	8.1	68.2	40	2.2
Chad	1978	4.3	83.6	56	2.0
Ethiopia	1976	29.7	87.7	65	16.7
Ghana	1978	11.0	64.0	55	3.9
Ivory Coast	1985	10.1	55.0	26.4	1.4
Kenya	1978	15.4	86.7	50	6.5
Lesotho	1979	1.3	95.7	55	0.7
Madagascar	1977	8.0	83.0	50	3.3
Malawi	1977	5.5	74.5	85	3.5
Mali	1975	6.3	82.8	48	2.4
Mauritius	1981	1.0	48.0	12	0.1
Niger	1975	4.7	89.7	35	1.4
Nigeria	1985	99.5	70.0	58	40.2
Rwanda	1975	4.5	96.3	90	3.6
Sierra Leone	1979	3.2	76.1	65	1.7
Tanzania	1978	17.6	98.2	60	9.0
Zaire	1975	24.7	65.2	80	12.9
Zambia	1975	5.0	30.0	52	0.8
Asia					
Bangladesh	1975/8	81.0	89.9	74–81	56.5
Burma	1978	33.1	73.8	40	9.9
China	1981	980.0	87	6–11	~ 60.0

Appendix A Continued

Country	Year of estimate	Total population (millions)	Proportion of rural population to total (%)	Percentage of rural population in absolute poverty	Estimated number of rural poor (millions)
India	1979	674.7	78.0	50.7	265.6
Nepal	1977	13.6	95.4	61	7.7
Pakistan	1979	84.5	72.2	39	23.8
Indonesia	1980	151.0	79.7	44	52.0
Korea Rep.	1980	39.0	40.0	9.8	1.6
Malaysia	1980–82	14.2	70.2	38	4.0
Papua New Guinea	1979	3.1	73.9	75	1.7
Philippines	1980–82	49.5	63.3	41	13.1
Sri Lanka	1981	15.0	72.0	26	2.8
Thailand	1978	44.4	85.9	34	13.1
Latin America					
Argentina	1975	25.5	32.0	19	1.6
Brazil	1980	122.5	32.0	68	26.5
Bolivia	1975	4.9	69.6	85	2.9
Colombia	1980	26.0	30.0	67	5.2
Costa Rica	1980	2.3	56.0	34	0.4
Dominican Republic	1978	5.3	51.0	43	1.2
Ecuador	1980–82	8.3	54.7	65	2.9
El Salvador	1978	4.5	59.4	32	0.9
Guatemala	1977	6.6	62.2	25	1.0
Haiti	1977	5.4	76.6	78	3.2
Honduras	1978	3.4	65.9	55	1.2
Jamaica	1982	2.1	50.0	51	0.6
Mexico	1982	73.1	52.0	34	12.8
Nicaragua	1978	2.6	47.9	19	0.2
Panama	1978	1.9	47.4	30	0.3
Paraguay	1978	3.0	61.2	50	0.9
Peru	1977	16.5	43.0	68	4.8
Trinidad & Tobago	1977	1.0	78.7	39	0.3
Venezuela	1980	15.0	20.0	56	1.7
Middle East					
Afghanistan	1977	14.4	86.0	63	8.2
Egypt	1982	44.2	56.0	17.8	4.2
Iran	1976	34.5	58.0	38	7.6
Iraq	1976	11.5	38.0	15–20	~0.8

Country	Year of estimate	Total population (millions)	Proportion of rural population to total (%)	Percentage of rural population in absolute poverty	Estimated number of rural poor (millions)
Jordan	1979	2.8	44.4	17	0.2
Morocco	1979	19.4	60.1	45	5.3
Somalia	1982	4.5	71.3	60	2.1
Sudan	1982	20.0	79.6	70	9.9
Tunisia	1977	5.9	50.7	15	0.4
Turkey	1986	51.0	54.0	20	5.5
Yemen, Dem	1978	1.8	64.1	20	0.2
					767.3

Sources: *The Dynamics of Rural Poverty*, Table 1.1, Food and Agriculture Organization of the UN, Rome, 1986. This appendix is reproduced with authorization from the Publication Division of the FAO. Changes have been introduced to update data for the following countries: *Nigeria*, estimate of rural poverty was made by Paul Collier in *Poverty, Equity and Growth in Nigeria and Indonesia*, Oxford University Press, forthcoming; *Korea, Republic* estimate is cited in: Shin Dong Wang and Choi Yang Boo, *Alleviation of Rural Poverty in the Republic of Korea*, see text, Chapter 6; *Mexico*, *World Development Report*, 1982, World Bank: 83; *Egypt*, see text, Chapter 6. The following countries have been *added* to the FAO list: *Ivory Coast*, estimate was made by the World Bank, Living Standard Unit. This estimate refers to per capita expenditure, Table 4 in *Confronting Poverty in Developing Countries*, Working Paper No. 48, 1988; *China* and *Iraq*, see text, Chapter 6; *Turkey*, see source in Table 5.4, Chapter 5.

Appendix B

The September 2000 declared millennium development goals for halving poverty and hunger by 2015

In September 2000, the heads of 189 states attending the Millennium Summit at the United Nations Headquarters, New York, adopted the Millennium Declaration specifying eight goals on several aspects of development. We are concerned here with the first goal on the eradication of poverty and hunger or chronic malnourishment.

This goal has two targets. The first is to halve, between 1990 and 2015, the proportion of people whose income is less than US$1 a day per person. The second target is to halve, between 1990 and 2015, the proportion of people who suffer from hunger. We have defined, in Chapter 1, chronic malnutrition or undernutrition meaning hunger, as defined by the experts of FAO and WHO.

Despite the specification of the proportion, not an absolute number, of poor people and in spite of the lack of impartial measurements (for example, it uses the controversial World Bank's measurement of US$1 a day that I and others have already criticized in Chapter 1), the goal is a tool for public monitoring of progress. It enables the non-governmental organizations (NGOs) and other civil society groups representing the rural poor to hold their governments and political leaders accountable. Moreover, the goal places poverty and chronic malnourishment at the centre of world and national debates on development. For example, both the March 2002 International Conference on Financing Development held in Monterrey, Mexico, and the September 2002 World Summit on Sustainable Development, held in Johannesburg, South Africa, backed the goal and reaffirmed the commitments by rich countries to increase financing for development.

For the purpose of its follow-up and the monitoring of progress made, the United Nations established the Millennium Project, comprising experts on the subjects of the eight goals. An important task to be undertaken is to assist poorer countries in building their statistical capacity, and to train national staff in charge of monitoring the changes in poverty and hunger, so defined. National teams would also need the support and cooperation of other country teams and international agencies' experts. This is not an easy task according to my own 5 years' experience within FAO for monitoring progress in the quantitative targets set by the 1979 World Conference on Agrarian Reform and Rural Development (WEARRD).

For further information, see:

1 UNDP (2003) *Human Development Report 2003, Millennium Development Goals: A Compact among Nations to End Human Poverty*, New York: Oxford University Press.
2 United Nations (2000) *Millennium Declaration*, press release, RES/55, September 2000.
3 United Nations (2002) *Millennium Development Goals: Data and Trends*, prepared by a UN group of experts on MDG indicators, New York: UN.
4 United Nations (2002) *Least Developed Countries Report: Escaping The Poverty Trap*, Geneva: UN.

For a critical assessment, see White, H. and Black, R. (eds) (2003) *Targeting Development, Critical Perspective on the Millennium Development Goals*, London: Routledge.

Appendix C
Summary of the 2006 land reform policy in Bolivia

Shortly after his election, President Morales announced, in May 2006, a new policy for the redistribution of nearly 2 million hectares of private agricultural land representing 65 per cent of the total arable land. This is in addition to an area of about 2 million hectares in forest and pasture land, the property title of which is being examined. The total of 4 million hectares is in the Departmentos (provinces) of Santa Cruz, Beni, Pando and La Paz. The intended beneficiaries are primarily the indigenous rural people (Andean Indian). The details of the programme are being discussed and worked out by the Comision Agraria Nacional comprising representatives of the government, the indigenous communities and the would-be affected commercial plantations. The outcome has to be approved by the parliament.

Earlier in 1953, a redistributive land reform was implemented in response to longstanding demands for restitution of landed property rights by the dispossessed indigenous *indios* and other landless peasants. A ceiling of 800 hectares in the rainfed highland *Altiplano* and 80 hectares in the fertile valleys was fixed in 1953, and the area above this ceiling was redistributed by the government in units of 10–20 hectares in the rainfed areas and 3–5 hectares in the valleys. In October 1959, the present author visited Bolivia and studied the implementation of the reform programme in the province of Cochabamba. I learned that the Peasants' Union, with the help of the Mining Workers' Union, was instrumental in the rapid implementation of the reform. I also found that no marketing arrangements for the beneficiaries were made to replace those previously monopolized by the former landlords in a highly controlled market network. Later, Clark (1968, 1970) studied the productivity in the beneficiaries' units and the farm management of the affected large landowners who were absentee and lived in the capital city, La Paz, while their farms were operated by bonded labour and managed by employed agents, particularly in the provinces of Santa Cruz and Oriente. De Janvry adds another disadvantage of the beneficiaries: 'All sources agree that the small holders, both within and outside the reform sector, have received virtually no production credit' (de Janvry 1981: 215–16).

The 2006 land reform policy, once enacted, will be in addition to that of 1953, which redistributed 18 per cent of the total arable land to 39 per cent of total agricultural households in 1956. Once the 2006 reform is implemented, Bolivia will have a complete land reform with the redistribution of nearly 84 per cent of total agricultural land. This anti-poverty, large-scale redistribution is opportune, considering the very high poverty level in rural Bolivia, reaching 77.3 per cent of the total rural population. Thus, Bolivia is one of the poorest developing countries (World Bank 2006: Table A1).

For further information on the 2006 land reform policy, visit the Bolivian website in Spanish at www.inra.gov.bo or contact the Embassy of Bolivia in London (email: info@embassyofbolivia.co.uk).

For references on the 1953 land reform and public actions taken since then, see:

Clark, R.J. (1968) 'Land reform and peasant market participation in the north highlands of Bolivia', *Land Economics*, May, 153–72.

—— (1970) *Land Reform in Bolivia*, Spring Review, Washington, DC: USAID.

De Janvry, A. (1981) *The Agrarian Question and Reformism in Latin America*, Baltimore, MD: Johns Hopkins University Press.

Bolivian Institute of Agrarian Reform: www.inra.gov.bo

Notes

1 The crisis

1 'LDCs' refers throughout this volume to the traditional developing countries before the World Bank's added transition countries (former Yugoslavia, ex-communist countries of Eastern Europe and the members of the former Soviet Union).

2 The developing countries with a prevalence of absolute rural poverty below 30 per cent of the total rural population are China, South Korea, Mauritius, Tunisia, Jordan, Egypt, Argentina, Nicaragua, Turkey, former South Yemen, Guatemala, Sri Lanka, Iraq and Ivory Coast.

3 These countries are in North Africa and the Near East region. My study was published by IFAD in 1990 with the title *Land Reform and Rural Poverty in the Near East and North Africa*, The State of World Rural Poverty Working Paper No. 22, Rome.

4 To measure poverty in the 114 developing countries, IFAD used an integrated poverty index calculated by combining the headcount ratio (simple proportion of rural population below the poverty line) with the income gap ratio, the distribution of income among the poor and the annual rate of growth of GNP per capita. In this case, the income gap ratio for each country is the difference between the highest GNP per capita from among the 114 developing countries and the GNP per capita of each country, expressed as a percentage of the former. [For detailed information on this index, see IFAD (1992) *The State of World Rural Poverty*: Technical notes to Table 2, 460–2.]

5 Detailed classification of the rural poor in Bangladesh in 1989 shows that the poverty headcount among the landless class of holding (0–0.04 acres) is the highest level in rural areas at 61.4 per cent; those in the near-landless class of 0.05–0.49 acres accounted for 53.9 per cent. The share of these two poorest groups was 13.9 and 31.5 per cent of total rural population (IFAD 2001: Chapter 3, Table 3.1).

6 BMR stands for basal metabolic rate, which is an estimated minimum dietary energy requirement using a safe range within which individuals can adjust their body weight in response to changes in energy intake without danger to health. The range is $1.2 \times$ BMR to $1.4 \times$ BMR. The difference represents a 7 per cent variation for adults of the same age and weight as empirically observed (FAO 1985a: 19–21).

7 See for example FAO (2003a: Chapter 8); Alderman *et al.* (2003); FAO/WHO (2004).

8 Sweden, Norway, Finland and the Netherlands have maintained their development assistance at more than the target of 0.7 per cent of their national income, and directed their aid to poverty reduction activities.

2 Reform of land property rights between the state and the market

1 For a concise statement on the operational significance of land tenure in its relationships with income distribution and productivity in agriculture, see Dorner (1964). On

land tenure and stages of development in a historical context, see Parsons (1962). Whereas water rights for irrigation are a crucial determinant of productivity and income for small farmers in arid and semi-arid areas, they have, nevertheless, received little analytical attention by either economists or technicians concerned with irrigation problems. The implications of institutional arrangements constitute the subject of a study prepared by Daniel W. Bromley for the World Bank (1982). It presents examples from Mexico, the Gezira scheme in the Sudan, Pakistan, the Philippines and Taiwan. For examples from other countries, see the bibliography of this study.

2 Dan Usher (1981: 85, 89) says: 'There is no capitalist equity without security of property [. . .] an economy cannot be said to be capitalist at all, if property is insecure, for the squabble over the assignment of property is every bit as corrosive to democracy as the squabble over the assignment of income. It makes no difference whether people fight over the fruit of the tree. With property, as with income, a degree of feasibility may be traded off for efficiency or for a greater acceptability of the system as a whole. But there are limits beyond which the security and rights of property cannot be attenuated [. . .] The legislature (by majority vote) has to specify the rules [. . .] and the line between the rights of property and the right each man enjoys by virtue of his status as a citizen'.

3 These Nobel Laureates for their contributions to the fields of institutional economics are (in chronological order with the year of award of the prize in brackets): Gunar Myrdal (1973), Herbert Simon (1978), George Stigler (1982), Raymond Coase (1991), Douglas North (1993). Joseph Stiglitz, George Akerlof and Robert Spence shared the prize in 2001, and Clive Granger and Robert Engle shared the 2003 prize.

4 See the study of 18 Egyptian villages in Radwan (1977) and El-Ghonemy (1986) concluding that inheritance is the dominant method of landownership in Egypt and Yemen.

5 The quotation is from Milton Friedman (1962: 161–2), and the American philosopher is David Haslett (1994: Ch. 6).

6 This information is taken from Scheftel (1947). This paragraph is based on a personal communication with the office of the Chief Rabbi, London (October 1993).

7 This principle has been clarified with Dr Hussein Hamed, the President of the International Islamic University, Islamabad, Pakistan.

8 For a comparative analysis of the ideas and approaches of the moral philosophers and the economists, see Roemer (1996: 2–4).

9 Just as the Norman Conquest of Britain in 1066 originated the granting of land by the sovereign and the feudal system, which was gradually reformed between 1286 and 1700, the pre-land reform concentration of large properties in Latin America, many South-East Asian and Middle Eastern countries resulted from grants during colonial rule. History tells us that British Viceroys, the Spanish Crown and Ottoman Sultans granted large estates to holders of certain offices and influential families on whose support the colonial rulers were dependent. These estates were legitimized and became hereditary possessions.

10 For a discussion of these elements as applied to pilot-scale projects, see El-Ghonemy (1999, 2002) and Carter (2000). See also de Janvry *et al.* (2002). In this standardized policy prescription, the evidence shows that the assumed linkage between free trade and high economic growth is inconclusive and doubtful (Wade 1990: 15–19; Singer 1992). From his careful analysis of LDCs' experience, Singer challenges the standardized policy by proving that poor countries are vulnerable to economic instability if they adopt the prescribed free trade regime in an imperfect global market.

11 Michael Lipton (1985: 4) suggests a significant relationship through his use of the multiplier effect and the accumulation effect. A landholder's household converts calorie intake via labour into an income level sufficient to satisfy dietary needs and to leave a surplus for accumulating new assets that reduce the future risks of poverty; he says, 'productive land turns out to be a much more life-cycle asset than income'.

12   For a comprehensive analysis of the effects of changing distribution on growth, see (i) Chenery (1979: Ch. 11); (ii) Chenery *et al.* (1981: Ch. 1 and 11); (iii) Fields (1980); and (iv) a careful review and analysis of the findings on the compelling evidence of the importance of distribution to economic growth and poverty reduction by Naschold (2004). For an understanding of the role of the imperfect rural labour market in generating poverty, see Collier and Lal (1986: Ch. 5 and 8).

13   On the demographic characteristics of the rural poor, see Visaria (1981), Lipton (1985) and FAO (1986: Ch. 2).

14   For a detailed discussion and critical review of the debate on welfare and utility theories, see Sen (1982: Parts III and IV). For a brief discussion on measuring inequality, see Bigsten (1983: 46–50).

15   The two measurements concern the comparison of changes in the degree of inequality. In this case, it is the share of size groups of landowners or landholders in the total area (e.g. 70 per cent of the owners own 20 per cent of the cultivated land). A Lorenz diagram is obtained by plotting the cumulative numbers. The extent of the inequality is indicated by the area between the diagonal (absolute equality) and the actual curve; the greater this area, the greater the inequality. The measure (Gini coefficient) must lie between 0 and 1. For example, an index of 0.876 is an indication of higher inequality of distribution than an index of 0.543. The Gini coefficient is a simple and direct measure for ranking inequality. The Theil index gives the same ranking as the Gini index when the curves do *not* intersect. The advantage of the Theil index is in measuring inequality *between* size categories and *within* each category of landowners or income groups.

16   These principles are concerned with how food demand depends on the size of income and food share in total expenditure. Keynes' marginal propensity to consume is the ratio of a small change in consumption to a small change in income. The short-term increase in the income of a poor man will not change his propensity to consume (average). Colin Clark collected data from previous research on food consumption at different levels of real income. The best measure is by converting the monetary terms into kg grain equivalent/person/year. He found that, at the lowest level of income, elasticity of demand for food measures the percentage change in food consumption with respect to a percentage change in income. Marginal propensity to consume, on the other hand, measures the *slope* of the demand line drawn through the points observed (see Clark and Haswell 1964: Chapter VIII, Chart XI, Table XXVII).

17   Mellor and Johnston (1984) report the results of country studies showing that, in India, the rural consumption of locally manufactured consumer goods is two-and-a-half times that of the urban consumption of those goods: 'In Asia, peasant farmers typically spend some 40 per cent of increments to income on locally produced non-agricultural goods and services. The income multipliers are substantial – in the order of 0.7, and the employment multipliers are probably larger.'

3 The agricultural dimension of poverty

1   Gross fixed capital formation (GFCF) is a valuation of investment in land improvement, soil conservation, land reclamation, irrigation, drainage, farm buildings and equipment. GFCF in agriculture is given for some developing countries in the United Nations *National Account Statistics*, which is issued annually.

2   By calculating capital/output ratio at 4:1 and the planned annual rate of agricultural gross domestic product (GDP) growth at 4 per cent, the share of agriculture in GDP at 20 per cent and the share of domestic savings at 15 per cent of GDP. Capital/output ratio is widely referred to in the literature on investment and is used in development planning. Nevertheless, it is arbitrary and ambiguous. Paul Streeten (1972) examined the limitations in its use: aggregation of diverse forms of capital as a homogeneous quantity; strong assumptions used with regard to its separation from human capital, institutions, management and organizations, attitude to work which influences output;

and problems of measurement (valuation of capital and output due to changing levels of prices), time lag between investment and output yielded, confusion between using average and incremental ratios, assumptions about rate of growth of the labour force, use of public expenditure on health, education and nutrition as consumption, etc. Streeten concludes that, given these difficulties and ambiguities, its usefulness in development planning is doubtful. See Streeten (1972: Ch. 6). However, based on Szepanik's study of 17 developing countries, Lipton (1977: Table 8.1) found that, in 11 of them, agriculture has fallen far short of adequate investment.

3   See El-Ghonemy (1998: 55) and World Bank (1995) *From Scarcity to Security: Averting a water crisis in the Middle East and North Africa.*

4   See Chapter 2 for the meaning of institutional monopoly in agrarian structure, land concentration and the Gini index.

5   I am grateful to the Statistics Division of the FAO for providing me with the results of the census from 15 countries; the rest of the total of 23 have been compiled and calculated by the author. The results of the Bangladesh census 1983–84 were obtained by the author from Volume I published by the Bangladesh Bureau of Statistics, May 1986.

6   On individualization of communal land in Africa south of the Sahara, see Platteau (1995) and Chapter 7 in the present volume.

7   See, for example, Alesina and Rodrik (1994), Perotti (1992), Birdsall and Londono (1997), Deininger and Olinto (2002) and Iradian (2005).

4   Farm size and productivity

1   See, for example, Backman and Christensen (1967), with comments by Heady in Chapter 7; Sen (1984: 37–72); Berry and Cline (1979: Ch. 2); Currie (1981) and Dandekar (1962).

2   Backman and Christensen (1967: 242).

3   In 1964, Professor Theodore W. Schultz examined the status of the operators of US farms between 1930 and 1960, and he classified them as owner-operators, tenants and hired managers. He found that there was a positive correlation between owner-operatorship and the rise in farm output per man-hour, i.e. the higher the percentage of owner-operators, the higher the output. Conversely, the higher the percentage of tenants and hired managers, the lower the output. He concluded that 'Absentee arrangements are in general inefficient'. He added that, when the market approach is adopted, 'The difference in the efficiency of absentee and resident production decisions in farming becomes relevant' (Schultz 1964: 120, 118, 104).

4   With the exception of Mozambique and Cuba, the data are taken from Kifle (1983). The data on Mozambique are based on my field visit in May 1984 (see *Report of the WCARRD Follow-Up Mission to Mozambique*, FAO, Rome, 1985). The data on Cuba are from an ILO paper by Peter Peek (1984: Table 6). The share of other farm institutions in agricultural land is 8 per cent for collective farming and 7 per cent for private farms. Most of the private farms are about 2 *caballerias* (28 hectares) in size. Between 1963 and 1983, the percentage of private farms to total agricultural land declined from 38 per cent to 7 per cent, whereas that of state farms increased from 61 per cent to 85 per cent (see Cuba in Chapter 5).

5   In alphabetical order, the countries visited/studied are: Cuba, Ethiopia, former East Germany, Hungary, Romania, Russia, former South Yemen and Yugoslavia.

6   This quotation should be seen against the background material on which it was based. Griffin admits that these measures are crude and should be regarded as rough approximations only. See Griffin (1986: 173, Tables 11.1 and 11.3).

7   These available cross-sectional studies are:

a   Barraclough (1973). Chapter Two presents the overall framework of the study. This volume was edited jointly with Juan Carlos Collarte.

b   Berry and Cline (1979). The questions addressed in this comprehensive and penetrating study are in Chapter 1. Chapter 5 presents the conclusions and policy implications.

c   Cornia (1985: 513–34). The analysis is based on data collected by the FAO Farm Management and Production Service between 1973 and 1979. For comparability purpose, all value figures were transformed into 1970 US dollars, and the area was converted into hectares. The countries covered by the analysis are: Sudan, Syria, Ethiopia, Nigeria, Tanzania, Uganda, Barbados, Mexico, Peru, Bangladesh, Burma, India, Nepal, South Korea and Thailand.

8   The following country-specific studies are listed in the same order as they appeared in the discussion:

a   Bhalla (1979). The sample consisted of some 3,000 cultivating households who were interviewed in each of the three years 1968/9, 1969/70 and 1970/1. Out of this sample, 1,772 were selected for analysis. Bhalla states that the oversampling of high-income households allows the data to be used for the study of the production behaviour of large (over 25 or 30 acres) farmers – a procedure not possible with most previous data sources in India. Also, its extensive coverage means that all the major crops grown in all regions of India can be analysed, and the different stages of adoption of the new technology can be meaningfully studied.

b   Berry and Cline (1979). The study of Brazil appears on pp. 44–58.

c   Hayami and Kikuchi (1985). This is based on case studies of two villages in the province of Laguna: the East Village and the South Village. Both are in a rice area. The two villages were surveyed in 1976 and 1977, and the analyses used an earlier survey of 1966 as a benchmark.

d   Henry (1986: 72–9).

e   Collier and Lal (1986).

9   There are many studies on technical change related to the size of farms. Examples are: Sen (1966); Azam (1973); Bardhan (1974); and Taussig (1982: 195). In his section on 'the green revolution in peasant agriculture 1970–72', Taussig describes, from his field surveys in 1970, 1972 and 1976, how all farmers adopted new varieties and inputs for growing soya and corn regardless of the size of holdings. The yields per unit of land were around no more than 60 per cent of those obtained on the large farms in the same area over 2 years 1970–72.

10   This remark refers to Cornia (1985). The countries studied are based on FAO data. For comparability purposes, see note 7c above.

5 Case studies I

1   I have not attempted to make a typology of land reform policy because I am convinced that it is controversial and overly rigid. There have been several attempts. For instance, Voelkner and French (1970) developed an analytical model encompassing all factors of social, political and economic change and development. They used 31 items to identify the historical phases of implementation. Each phase is related to the development stage reached by the country concerned. But their definition of land reform is very broad, and fits into the ambiguous term 'agrarian reform' explained earlier in Chapter 2 in this volume. On the other hand, de Janvry (1981) categorizes land reform policies according to the degree of change made in production and class relations. His broad classification of initial land reforms falls under semi-feudal, capitalist and social modes of production. According to this typology, de Janvry classifies 33 land reforms in 20 countries.

2   I measure the characteristics of this tendency in terms of the inequality index of the Gini coefficient of land and income or expenditure distribution illustrated by the Lorenz curve. If a tendency towards increased inequality in rural areas is evident, we attempt to identify the conditions that mark its turning point.

3 Most statistics used in the pre-reform period are taken from this comprehensive volume. It contains the results of Chinese scholars' research work in different provinces on subjects concerning the distribution of landed property, farm management, agricultural marketing and credit, and rural handicrafts supplementing the peasants' incomes from the land. The contents are reviewed by R.H. Towney, Professor of Economic History, London School of Economics, who emphasized that any realistic study of modern China must start from the question of land tenure. He stated that the writing of the Chinese scholars points to the fact that 'rural society in China has reached a crisis' (Institute of Pacific Relations 1939: xvii).

4 For a detailed study of why population growth is controlled in China, see Saith (1984).

5 This was based on a study prepared for 1977, 1978, 1979 and 1981 by the Commune Management Bureau of the Chinese Ministry of Agriculture. Chronic poverty is defined as per capita distributed collective income averaging less than 50 yuan in each of the three years 1977 to 1979. The results by localities are analysed by Lardy (1983: Ch. 4). The number of counties with chronic poverty declined from 221 in 1979 to 87 in 1981. Lardy argues that the Ministry's study underestimates poverty among teams within counties. The estimate of 60 per cent poverty before the reforms in the late 1940s is calculated from data for 'poor peasants' in the Chinese studies in Institute of Pacific Relations (1939). All cited estimates on rural poverty prevalence in China are approximate, showing only the order of magnitude.

6 An important source of widespread employment and earnings has been the expanded network of small-scale industries in all rural areas. In these enterprises, women have the same entitlements as men, the equipment is simple, the level of technology is low and the intensity of labour is high. (The concern for maximum utilization of labour has probably been behind the limited scale of the mechanization-based state farms to only 4 per cent of total cultivated land.)

7 The USAID data for 1945 cited in Table 5.5 refer to the size distribution of units as 'management scale', which can be interpreted as operational holdings. Eddy Lee says, 'Thus the main thrust of the land reform was the replacement of tenancy by owner cultivator and not a radical shift in the size distribution of production units (holdings). 60% of all cultivated land was tenant-farmed in 1945 and at least 50% of farm households are pure tenants' (Lee 1979: 25, 26).

8 *Saemaul Undong*, meaning new community movement, represents a significant feature of rural development in South Korea. It started in 1971 and is based on the voluntary participation of local rural people in creating small irrigation schemes for rice cultivation. It later developed into establishing small industries, rural electrification, the spread of high-yielding varieties, improvement of rural housing and providing safe drinking water. In addition, it trained young farmers, consolidated fragmented holdings and constructed rural roads. This movement gained strength in the 1970s and was instrumental in the relocation of industries in rural areas. On average, the funds used were made up of 70 per cent contributions from the rural people themselves and 30 per cent from the central government.

9 The average income in monetary terms was deflated by using an income group-specific index based on consumption weights. Eddy Lee examined the data of the Farm Households Income Surveys conducted by the Korean Ministry of Agriculture and compared the differential changes by size of holding. Average income of holders in the size group of less than 1 hectare (where most land reform beneficiaries lie) rose between 1963 and 1975 by 54 per cent, while that of holders of more than 2 hectares increased by 47 per cent. Agricultural wages in real terms rose from 167 won per day in 1963 to 237 won in 1975 (see Lee 1979: Tables 2.4 and 2.6).

10 Annual net outmigration from farm households rose from an average of 243,000 persons per year during the period 1960–66 to 568,000 persons per year during 1966–70. Over half the total migrants were between the ages of 15 and 30. This fast movement

resulted in a decline in the farm working-aged population in proportion and in absolute numbers (see Ban 1980: Ch. 12).

11 These projects were developed on state-owned land in Dujaila, Sinjar, Hawija, Latifiya, Mussayeb and Makhmur. The planned area for redistribution was 2,126,580 donums, but the area *actually* distributed was 232,960 donums according to Hassan Mohamad Ali, the Chairman of the Land Development Committee, who accompanied me on a study visit to Latifiya and Mussayeb during 1955. The present author also visited the Dujaila scheme with Dr Burnell West, who was working as the FAO soil expert in Iraq. It was noted that: (i) salinity had increased due to overirrigation, and had forced land out of production in many parts; (ii) about a quarter of the settlers were former civil servants with no experience in farming; and (iii) the approach in planning and implementation was an engineering and not a rural development approach. During his visit, the present author was preparing a study on land tenure and settlement for the FAO Regional Symposium on Land Problems in the Middle East, held in October–November, 1955, at Salahuddeen near Irbil in the north of Iraq.

12 I say roughly estimated because, during my visit in 1964, I was given conflicting statistics on land already expropriated and land subject to requisition by the Ministry of Agrarian Reform. The figures on the latter ranged from 6.3 to 10 million donums. With regard to compensation payments, the land reform law of May 1970 specifies compensation payments in the case of requisitioned, but not confiscated, land.

13 The author was invited twice by two Ministers of Agrarian Reform, Dr Abdul-Saheb Alwan in 1964 and Dr Ahmad Al-Dujaili in 1967, to give advice on specific issues in the course of land reform implementation. He also visited Iraq in 1975 to study the equity and production implications of the second land reform law of 1970.

14 When Colonel Abdul Karim Al-Qassim, the leader of the July 1958 Revolution, proclaimed land reform on 30 September 1958, he stated: 'We have found that agrarian reform is the foundation of social reform which our great nation is historically entitled to [. . .] the law will do away with feudalism forever and liberate the fellaheen'.

15 Landless and sharecroppers were estimated at 73 per cent of total agricultural households in 1952 and 67 per cent in 1958 from the results of the 1957 population census and agricultural censuses of 1952 and 1958.

16 See *Food Consumption Tables in the Middle East*, FAO, 1970; *Recommendations of FAO/WHO ad hoc Expert Committee*, 1973; and the *Fourth World Food Survey*, FAO, Rome, 1977. In these documents, the conversion is made on the basis that 315 calories equals 100 grams of wheat.

17 The subsample data of the 1972 and 1976 household surveys analysed by Bakir (1979) include a 'low-income' group in rural areas. Its average annual per capita expenditure in 1976 was 57.2 Iraqi dinars. Assuming that the expenditure of this group is almost equal to their incomes, their average expenditure amounts to only 15 per cent of the national average income (389 dinars at 1975 prices estimated by the Ministry of Planning). The low-income group is the only classification in rural areas that was at high risk of undernourishment. The nearest group is the rural south. These two groups, low income and rural south, were the only ones whose average annual per capita expenditure fell in real terms between 1972 and 1976; the former by 4.6 per cent from 40.2 to 40.1 dinars, and the latter by 8.7 per cent from 62.7 to 58.8 dinars (calculated from Bakir 1979: Tables C12, C23, C28 and 5.23). This decline is manifested in deterioration in 13 out of 24 indicators between 1972 and 1976.

18 The present author, in his capacity as land tenure specialist with the United Nations FAO, was a participant in and coordinator of an international team whose visit was requested by the Cuban Authority. The statement and observations made here were based on a 2-month (July and August 1959) survey of the agrarian system, providing advice on the economic basis for fixing the minimum units for allotment to the beneficiaries and giving his views on the implementation capability to the National Institute for Agrarian Reform (INRA). The team consisted of Professor V.M. Dan-

dekar, Gokhale Institute of Politics and Economics, Poona, India, Professor Marco Antonio Duran, Head of Agricultural Economics Department of the National Agricultural College at Chapingo, Mexico. See *Report of the Regional Land Reform Team for Latin America*, FAO, Rome, 1961, Expanded Programme of Technical Assistance, No. 1388.

19 Sugar cane harvest declined in this early phase of Cuban land reform primarily due to organizational and management shortcomings. It fell from 55.8 million tons in 1960/1 to 50.6 million tons in 1965. Since then, it has started rising and reached 82.9 million tons in 1970. The management shortcomings are frankly described by Ernesto Guevara (1964). See also Dumont (1964: 42–59).

20 *Ejido* is a Mexican term referring to an indigenous system of land tenure and use. It is communally owned and operated agricultural land. Most of their lands were gradually taken away by US capitalists and rich Spaniards. Hence, the Zapata revolution of 1910 prohibited landownership by foreigners, and the restitution of the lands lost from the *Ejidos* was built into the Mexican Constitution of 1917.

6 Case studies II

1 This estimate is based on a minimum household net income of E£35 in 1949/50. A breakdown of the database is as follows:

My estimated number of poor households

Hired agricultural workers over 15 years of age calculated from the results of the 1947 Population census and the 1949/50 Agricultural census: 1,060,150

Annual earning per working person was on average E£13–15 based on: (i) my own studies carried out during work in the Fellah Department, and (ii) the rural survey of 2,682 landless households. This latter study was carried out by the field staff of the department in 1950 under the guidance of Dr Zelenka and Dr Cassidy from ILO, Geneva, with whom I worked as their research assistant. The purpose was to estimate the number of households whose earnings were below the established minimum income of E£35 for social security purposes. See El-Ghonemy (1953: 72–3).

Sharecroppers (agricultural census of 1950): 6,900

Holders of less than 1 feddan (1950 agricultural census): 214,300

Annual average net value added per feddan was between E£8 and E£15 based on a study by the Ministry of Agriculture on *National Income from Agriculture* 1946–47, Cairo, Table 11, p. 41 (in Arabic).

Pure tenant cultivators of 1–2 feddans: 150,000

Calculated from the results of the 1950 Agricultural census. The income of tenants in 1947–50 was estimated by the present author; see El-Ghonemy (1953: 57–9).

Family heads of nomadic population camping around villages estimated by the author from observations in the districts of Delingat, Abu-Hommos and Korn Hamada in Bohera province in 1949: 30,000

Peddlers in villages, small towns and *Ezbas* (big farms): 25,000

Disabled heads of rural households 0.2% (a rate used in the population census that was close to the rate found in the 1950 study cited above): 30,140

Total number of estimated heads of poor households: 1,537,090

Number of poor persons (using uniform average of five per family): 7,685,450

Total rural population in 1950 (based on 1947 population census and average annual rates of growth between 1947 and 1950 censuses): 13,710,000

Estimated percentage of poor rural population: 56.1

2 A separate administration was set up, free of the government's rigid rules and headed by a very able agricultural expert, Mr Sayed Marei. He selected the best managers of the former royal estates to administer the implementation of land reform laws at the field level. This administration was supported by a tough member of the Revolutionary Council, Gamal Salem, who was given full authority to take necessary action across

the Egyptian government. At the time of implementing the reform, Egypt already had a complete set of land title records and cadastral maps, as well as a sufficient number of university graduates in agriculture, veterinary science and irrigation engineering. The implementation of the 1952 law was scheduled to be completed by 1958.

3 This state was created by the affected landlord practice of subdividing part of their large landownerships into units of 10–50 acres before expropriation. Article 4 of the Land Reform Law of 1952 (amended in 1953) allowed the owners of 200 feddans and more to transfer part of the excess (area over 200 feddans) to their children (50 feddans per child) provided that the total area transferred did not exceed 100 feddans. This meant that each family could retain 300 feddans legally. Through these private land property transactions, many landlords were able to escape forced expropriation of land in excess of 300 feddans.

4 These studies include Gadalla (1962), Warriner (1962), Saab (1967), El-Ghonemy (1968a), Abdel-Fadil (1975), Radwan (1977) and Hopkins (1987).

5 This survey was based on a 5 per cent random sample of beneficiary households in the three land reform cooperative areas.

6 Food subsidies have been a significant feature of the post-1952 government policy to support low-income groups. Although beginning with an urban bias, the benefits have spread in rural areas through the network of cooperatives selling subsidized commodities, particularly bread, wheat flour, sugar, tea and edible fats and oil for cooking. The government allocation for food subsidies dramatically increased with the population growth, from 0.3 per cent of total public expenditure in 1965/6 to 27 per cent in 1980, and from less than 0.1 per cent of GDP to 10 per cent in the same period. In addition, we estimate that about E£15 is gained per consumer per year from subsidization of other commodities (cloth, fuel and transport). This average of E£30 represents a nearly 20 per cent indirect increase in the real income of small farmers if they purchase the subsidized items. For a detailed discussion on the subject of food subsidies, see von Braun and de Haen (1983).

7 It is unfortunate that most of the available information is either guesswork or a crude estimate. For example, estimates of the total number of Egyptians working abroad range from 1 to 5 million in 1980–84. The same applies to the inflow of remittances. Official data provided by the Central Bank of Egypt underestimate the *actual* transfer of remittances entering Egypt every year. Official statistics tell us that remittances rose from E£212 million in 1975 to E£3.3 billion in 1984, a 16-fold increase. The Ministry of Labour in Cairo estimated the Egyptian landless labourers working in Arab states at 1 million in 1986 (*Al-Ahram*, 30 June 1986: 3).

8 Radwan and Lee arrived at different estimates by using different sizes of household and income corresponding to a poverty line that differed from that used by Adams. Their estimates of the rural poor were 22.5 per cent (1958/9) and 17 per cent (1964/5). The three scholars used the same source of data, i.e. the results of the Egyptian Government Household Budget surveys for 1958/9, 1964/5 and 1974/5. We did not use Adams' estimate for 1974/5 at 60.7 per cent because we consider it an overexaggeration using the poverty line at E£344.82, which is almost the overall average expenditure per rural household in that year.

9 The partial land reform has to be seen in the context of India's programme for rural development since the early 1960s, which has had several elements, including: (i) community development programmes under which 100 villages with about 65,000 persons were to constitute an administrative block with service cooperatives established in villages; (ii) the Intensive Area Development Programme, which replaced the former programme and which concentrated mostly on irrigated areas with a technological thrust for increasing the yields of wheat, using tractors and tube wells; (iii) credit supply programmes (small farmers development, marginal farmers and landless labourer programmes); (iv) the rapid expansion in irrigated land during the 1970s; (v) Integrated Rural Development Programmes; and (vi) a National Rural Employment

Programme. Among other national programmes benefiting the rural poor, the population growth control policy was targeted to reduce the fertility rates in the large family size groups, who are mostly the poor.

10 No attempt is made here to extensively review Indian land reform policy, on which there is a rich literature. Examples are: the periodical reports on 'Progress of Land Reform' prepared by the Land Reform Division of the Planning Commission; Rural Labour Enquiry 1974–75; Joshi (1961, 1975); Dandekar (1964); Warriner (1969); Bardhan (1970: 261–6).

11 I visited India twice, in 1964 and 1968, to conduct a field study of land reform programmes in West Bengal and the Gram Panchayat. In Delhi, I was informed by Professor Gadgil, the then Vice Chairman of the Planning Commission, on the overall progress and problems in land reform implementation. Very useful discussions with Professor Karve and Professor Dantwala were held, which enlightened me on the functioning of agricultural cooperatives. Subsequently, several discussions took place with Professor Parthasarthy (Andhra University) and Dr T.C. Varghese (FAO) on a wide range of land reform policy issues in specific states in India.

12 In addition to free education at the primary and secondary levels, the state government of Kerala provides social security payments and unemployment allowances as well as monthly payments to the disabled. According to official government statistics cited in Jose (1983), the average per capita expenditure on education was Rs 62.8, while the corresponding average for all India was Rs 33.7. The expenditure in 1978 on education represented 45 per cent of total expenditure in Kerala during the 1960s and 1970s.

13 This study was conducted for the FAO jointly with Professor Kenneth Parsons of the University of Wisconsin, USA, who was commissioned by USAID. The findings were presented at an FAO meeting on land policy held in Salahuddin in northern Iraq, in October 1955.

14 On the land tenure system, see Lampton (1958) and its second edition (1969). On the Iranian meaning of property in land, see Denman (1973: Ch. 2). Data on income distribution are taken from Issawi (1982: Table A.4).

15 Dr Arsanjani called for 'land to the tiller' and, when he become Minister of Agriculture in 1961, he efficiently completed the first phase of land reform. The US intervention and successful experience in land reform programmes in Japan, Taiwan, South Korea and southern Italy encouraged the American government to press for land reform in Iran, which had the then Soviet Union and its communist propaganda right at its northern borders. The late President Kennedy's Foreign Assistance Act of 1961 (Sections 102 and 103) stated that 'the establishment of more equitable and more secure land tenure arrangements is one means by which the productivity and income of the rural poor will be increased'.

16 During my 1961 and 1965 visits, I had discussions with, and collected information from, Dr Hoshang Ram, Director of the Bank Omran of the Pahlavi Foundation, Dr Webster Johnson and Mr Kenneth Platt of USAID, and Dr Price Gittinger, economist at the Plan Organization. I interviewed farmers in the villages of Khaveh, Toghan, Dawood Abad and Khare Abad, which had Crown land distribution programmes. I also visited Crown land villages in the northern region near the Caspian Sea, where army officers received 30–50 hectares each. Accompanied by Mr Mohammed Ja'far Behbahanian, the head of the Shah's Pahlavi estates, I attended the ceremony at which the Shah himself presented property titles to the new owners at the Marble Palace in Tehran on 8 April 1961. For the complete story of the programme, see Farhang Rad's account in El-Ghonemy (1967: 273–91).

17 See Amid (1990: 103) for the superior quality of land retained by landlords.

18 I was assisted in this village study by Mr Ahmed Behpour of the research centre of the Land Reform Agency.

19 Public investment favoured the corporations to the disadvantage of the peasant sector,

comprising the beneficiaries of land reform, who were new owners and many small tenants who gained rent reductions; their area represented 85–90 per cent of the total landholdings area in 1978, compared with only 3 per cent of the corporations. See Moghadam's study (1978) cited in Katouzian (1983).

20 The farm corporations that I visited were Aria Mehr (1,500 hectares), Saheb Granich (20,000 hectares managed by the army for the production of meat and caviar near Rasht on the Caspian coast) and Roke Beesh (710 hectares in Gilan province). The assessment in the text was based on my interviews with Hassan Askaripour. Field research conducted by Najafi (1970) in three farm corporations showed that 44 per cent of the farmers did agree to surrender their individual rights to farm corporations.

21 Shirazi (1993: 194, Table 8.2).

22 See Lazarev (1977).

23 This statement and the discussion that follows are based on an unpublished study that I prepared in 1989 for the United Nations ECA. The material is used here with permission from the Joint FAO/ECA Agriculture Division, Addis Ababa.

24 Using the official data on the distribution of land and those of the United States Agency for International Development (USAID), Swearingen shows how the pace of agrarian reform was associated with the Casablanca riots in March 1965 and the attempted coup by rebel military officers in July 1971 and August 1972 (1988: Table 13, 177).

25 These programmes include: Agricultural Tenancy Act, 1954; Agricultural Land Reform Code, Republican Act 3884 issued in 1963; land settlement schemes in Minariao and Banislan, where nearly 850,000 hectares was settled by former squatters on public land and landless households numbering about 50,000; associations for the beneficiaries of land reform called *Samahang Nayon*; rural banks for agricultural credit and the employment-generating programme known as *Kilusang Kibuhayan* (KKK) to help landless workers, forestry workers and poor fishermen.

26 To appreciate the complex nature of land reform implementation, we should note that the Philippines consists of 1,000 inhabited islands, stretching for 1,850 kilometres, and that only 32 per cent of the total area is suitable for agricultural purposes. The increasing farm population pressure on cultivable land is manifested in the results of the Census of Agriculture for 1960 and 1991. The results show that, while the number of landholdings (or farms) increased from 2.17 million in 1960 to 4.61 million in 1961, their total area increased from 7.7 million hectares to 9.97 million hectares, i.e. an increase of nearly 112 per cent by number but only 28 per cent by area. Alarmingly, the number of very small landholdings of less than 1 hectare increased from 10 per cent in 1960 to 33 per cent of the total number of holdings. For a recent study on population pressure and land tenure, see Guiltiano *et al.* (2003).

27 There are many who are critics of the Philippines' limited land reforms and their slow implementation and who have been demanding comprehensive land reform. They include Mangahas, Montemayor, Po, Ledesma, Umali and Quison, just to mention a few. In addition, the numerous non-government organizations have consolidated their efforts during 1986–88 and voiced the misery of landless workers. After the announcement of the Presidential Executive order on land reform in May 1987, the pressure has increased for a land-to-the-tiller reform abolishing absentee landownership and covering all lands irrespective of crop planted. A summary of the work of these NGOs towards land reform is published in *Information Notes* (Vol. 8 No. 6, May 1988) by the Asian NGO Coalition for Agrarian Reform and Rural Development (ANGOC) in Manila. In 1998, President Ramos extended the Aquino comprehensive programme until 2008 (see Guardian 2003: 71–82).

28 Since 1957, there have been several estimates of rural poverty derived from different poverty lines. These estimates were made by the World Bank, the Philippines Wage Commission, Philippines Development Academy and Dr Mangahas. The methodological problems in estimation are discussed in Mangahas (1985), the World Bank

study on the Philippines (1980) and in Technical Appendix I of El-Ghonemy (1986). All estimates (except that of Mangahas) are based on data given by the Family Income and Expenditure Survey conducted by the National Census and Statistics Office in 1956, 1961, 1971 and 1975. The variation in estimation is due to using different percentages of food expenditure to total, per capita or household consumption per year, different pricing of the food and non-food items and inter-regional variation in consumer price indices and their adjustment to different rates of inflation. The result is a divergence of poverty lines for a rural household of six persons for 1975 as follows:

World Bank   4,962 pesos
The Academy   8,668 pesos
The Wage Commission 6,900 pesos

29 A recent example of this rainfall instability is the 1999 drought and its prolonged negative effects, which continued into 2000 and, according to my estimate, wiped out nearly 14 per cent of total agriculture value added, comprising 40 per cent loss in rainfed products and 64 per cent in total wheat production. This fall in agricultural output in wiped out nearly 8.4 per cent of Syria's GDP in 1999/2000 (see El-Ghonemy 2005).

30 Calculated from Syria's *Statistical Abstract* (2000, 2003).

7 Market liberalization

1 For an account of the aims and perceptions of the US government with regard to land tenure problems in developing countries and the views of several scholars and senior administrators, see the conference proceedings edited by Parsons *et al.* (1956).

2 This quotation is from President Kennedy's speech at the Alliance for Progress Conference in Punta del East, Uruguay, in 1961.

3 In its *Focus on Poverty* (1983), the World Bank admits that rural development projects did not reach all the poor and that they have provided few direct benefits for the landless, for tenants unable to offer collateral for loans and for the near-landless farmers who find it hard to borrow and acquire inputs.

4 For a detailed analysis of field studies on the contrast between the assumed marketability of land in theory and the realities in Egypt and India – apart from leasing – in land and distress sales by peasants, see El-Ghonemy (1992), Radwan and Lee (1986) for conditions in Egypt, and Bharadwaj (1985) and Harriss-White (1996) for the situation in India.

5 The international agencies mentioned in the text are: the Inter-American Development Bank, the World Bank, the FAO of the UN, IFAD, UNDP, UNESCO, the European Union, the USAID and the United Nations Commission for Latin America and the Caribbean.

6 Government of South Africa (1996).

7 Government of South Africa (1996).

8 Martin Adams (2000: Table 7) lists the donors and their substantial aid to the land reform policy in the Overseas Development Institute's study as follows: European Union, €11.2 million; UK, £5.5 million; Ford Foundation, US$0.9 million, in addition to several instalments paid by the Netherlands, Denmark and Switzerland.

9 The National Land Committee consists of 10 NGOs.

10 FAO (1993a) and FAO (1997a). Between 1989–91 and 1997, the tobacco area increased by 13,000 hectares, while that of maize fell by 92,000 hectares.

11 On a recent assessment of poverty in Egypt, see El-Leithy *et al.* (2003), a World Bank and Egypt Ministry of Planning joint study. It shows that the poor are concentrated in agriculture (57 per cent of the total).

12 I conducted a statistical analysis to estimate the relationship between rental values as dependent variable and agricultural output value together with cotton price and the land concentration index as independent variables. The hypothesis is that cotton

price, rent and output are positively correlated in a private property market economy and that cotton price is a determinant of the profitability of holding land. It is assumed that technical change is captured by variation in agricultural output value. The results of the analysis show that, during the first period, the variables have the correct (expected) sign, i.e. the higher the rental values, the higher are the output value and cotton price ($R^2 = 0.7$). Only cotton price and land concentration and not output value are determinants in the second period. For the entire period 1913–86, $R^2 = 0.9$, and the method of estimation is ordinary least square. See the equations on the results in El-Ghonemy (1992).

13 For an account of the Philippines land reform performance and the problems encountered, leading to the adoption of the market-based approach, see Putzel (2002). See also the complete proceedings of the *Colloquium on Agrarian Reform* (Government of the Philippines 1990), in which a concise statement on the problems encountered is presented by Dr Cornista, the Director of the Institute of Agrarian Studies (pp. 13–21).

14 For a detailed study on privatization of land and decollectivization of agriculture, see the work carried out by a team of researchers over the period 1991–95 (Szelenyi 1998).

8 Challenges and prospects

1   It could be argued that the distribution of landownership – and not landholdings – is more relevant to questions of poverty and income inequality. On the other hand, the fact that many landowners rent out part or all of their landholding means that there are several landholders of a single landownership unit. Therefore, the FAO, in its design of international agricultural censuses, uses 'landholding' or operational unit or farm as the unit of enumeration.

2   In El-Ghonemy *et al.* (1993b: 360), linear, semi-logarithmic and double logarithmic equations were fitted by ordinary least squares. The best-fit equation is

$$P = 6.194 - 0.274X + 1.65G$$

$$R^2 = 0.70 \qquad n = 21$$

where P = headcount ratio of rural poverty, X = agricultural GDP per agricultural labourer and G = Gini index of land distribution inequality.

3   For a detailed comparison between all India and the state of Kerala over the period 1977–2000, see Chapter 6, Table 6.8, and Sen (2002: Tables 9, 10, A6b and A6c).

4   Arable or cultivable land is the amount of land available for farming, irrespective of its intensive use. If cropped twice a year, the land intensity ratio is almost 2, so 1 acre or hectare of cultivable land is counted as 2 acres or hectares of cropped land.

5   In El-Ghonemy (1993a: 68–9), I reported that the rich farmers (50 hectares and over in Morocco), representing only 0.5 per cent of the total number of landholders, captured a disproportionately high share of 43 per cent of total government-subsidized production inputs. Their share in subsidies was even higher (at 85 per cent) in tractors and combine harvesters.

6   See note 2 above.

7   On the working of government-established institutional arrangements to enable the market to function effectively in South Korea, Japan and Taiwan, see Johnson (1987) and Wade (1990).

8   In Pakistan, the collective power of tenants has succeeded in withholding old rents once control was legally instituted. As an organized group, they have been able to hire lawyers to take their case through complex court proceedings and demand their rights.

Such power is effective in the face of legally prescribed reforms which remain unenforced by local bureaucracies under unaltered political systems (see Rahman 1983).

9　The right of agricultural workers to organize is regulated by international conventions monitored by the International Labour Organization (ILO). Important among these instruments are Conventions 87 and 141 on the freedom of association and protection of the right of rural workers. Even after official ratification, it was reported that some countries violated their provisions and departed from their principles. See a recent account in UNDP (2003: Indicators Tables).

10　For a detailed account of this landless workers' movement, see Wendy Wolford, 'Case Study: Grass roots – initiated Land Reform in Brazil, in de Janvry *et al.* (2002). Also in the Philippines, nearly 20 NGOs with support from church leaders have been actively lobbying for a greater access to landownership. In May 1987, they formed the Congress for People's Agrarian Reform to lobby collectively for what they called 'genuine agrarian reform'. Their efforts were coordinated by the Asian NGO Coalition (ANGOC) under the leadership of Dr Umali, formerly the Dean of the College of Agriculture, Los Banos. They were joined by church leaders and a number of scholars. This consolidated effort produced a plan calling for land to the tiller; complete abolition of absentee landownership and coverage of all crops and tenure arrangements. On the efforts of NGOs in the field of land reform, see Ghimire and Moore (2001). On the international level, both IFAD and UNRISD have effectively supported the Popular Coalition to Eradicate Hunger and Poverty, which co-sponsored the preceding publication of 2001.

Bibliography

Abate, A. and Kiros, F. (1983) 'Agrarian Reform, Structural Changes and Rural Development in Ethiopia', in Ghose, A.K. (ed.) *Agrarian Reform in Contemporary Developing Countries*, London: Croom Helm.

Abdalla, A. (1993) *Formal and Informal Organization of Agricultural Land Markets in the Sudan*, Rome: FAO.

Abdel-Fadil, M. (1975) *Development, Income Distribution and Social Change in Rural Egypt*, Cambridge: Cambridge University Press.

Abu-Zahra, Mohammed (1963) *Ahkam al-tar' ikaat wal Mawareeth* (*Laws of Inheritance and Legacies*), Cairo: Dar al-Fikr al-Arabi.

Adams, D. (1958) *Iraq's People and Resources*, University of California.

Adams, Martin (1997) *The Importance of Land Tenure to Poverty Eradication and Sustainable Development in Africa*, Oxford: Oxford Management Policy.

—— (2000) *Breaking Grounds: Development Aid for Land Reform*, ODI Research Study. London: ODI.

Adams Jr, Richard (1985) 'Development and Structural Change in Rural Egypt, 1952–1982', *World Development*, 13, 705–23.

—— (2003) 'Evaluating Development in Egypt', in El-Ghonemy, M. (ed.) *Egypt in the Twenty-First Century*, London: Routledge.

Adelman, I. (1974) 'South Korea', in Chenery, H., Ahluwalia, M., Bell, B., Duloy, J. and Jolly, R. (eds) *Redistribution with Growth*, Oxford: Oxford University Press.

—— (1987) *Practical Approaches to Development Planning: The Case of South Korea*, Baltimore, MD: Johns Hopkins University Press.

Ahluwalia, M. (1985) 'Rural poverty, agricultural production and prices', in Mellor, S. and Desai, G. (eds) *Agricultural Change and Rural Poverty*, Baltimore, MD: Johns Hopkins University Press.

Aiguo, Lu (1996) *Welfare Changes in China during the Economic Reform*. Research for Action No. 26, Helsinki: UNU/WIDER.

Alamgir, M. (1977, 1980) *Famine in South Asia – Political Economy of Mass Starvation in Bangladesh*, Dacca: Bangladesh Institute of Development Studies.

Alderman, H., Hoddinott, J. and Kinsey, B. (2003) *Long-Term Consequences of Early Childhood Malnutrition*, Washington, DC: Institute of Food Policy Research.

Alesina, A. and Rodrik, D. (1994) 'Distributive Politics and Economic Growth', *The Quarterly Journal of Economics*, 198, 465–90.

Alex, J. and Almeda, J. (1980) *Wage Rates of Farm Workers in the Philippines*, Manila: Bureau of Agricultural Economics.

Alexandratos, Nikos (ed.) (1988) *World Agriculture Towards 2000*. An FAO Study, London: Pinter, Belhaven Press.

Ali, Hassan (1955) *Land Reclamation and Settlement in Iraq*, Baghdad: Majlis al-Imar.

Ali, M.S. (1985) 'Rural poverty and anti-poverty policies in Pakistan', in Islam, R. (ed.) *Strategies for Alleviating Poverty in Rural Asia*, Bangkok: ILO.

Alwan, A.S. (1985) *Agrarian Systems and the Alleviation of Poverty in Iraq*, Baghdad: UN Economic and Social Commission for Western Asia (ESCWA).

Amid, M. (1990) *Agriculture, Poverty and Reform in Iran*, London: Routledge.

Arcand, Jean-Louis (2001) *Undernourishment and Economic Growth: the Efficiency Cost of Hunger*. Economic and Social Development Paper No. 147, Rome: FAO.

Arodki, Y. (1972) *Al-Iqtisad Al-Soury Al-Hadith (The Recent Syrian Economy)*, Damascus: Al-Tarabishy Publishing House.

Atkinson, A.B. (1975) *The Economics of Inequality*, Oxford: Oxford University Press.

—— (1983) *Social Justice and Public Policy*, Cambridge, MA: MIT Press.

Assembajjwe, G., Banana, A. and Bahati, J. (2002) 'Case Study: Property Rights in Uganda', in de Janvry, A., Gordillo, G., Platteau, J. and Sadoulet, E. (eds) *Access to Land, Rural Poverty and Public Action*, Oxford: Oxford University Press.

Azam, K. (1973) 'The Future of the Green Revolution in West Pakistan', *International Journal of Agrarian Affairs*, 5(6), March.

Aziz, S. (1978) *Rural Development: Learning from China*, London: Macmillan.

Backman, K. and Christensen, R. (1967) *The Economics of Farm Size: Agricultural Development and Economic Growth*, Ithaca, NY: Cornell University Press.

Bakir, M. (1979) 'The Development of Level of Living in Iraq' unpublished PhD Thesis, University of Leeds, UK.

Ban, Sung-huan (1980) *Rural Development*, Cambridge, MA: Harvard University Press.

Baranzini, M. (1991) *A Theory of Wealth Distribution and Accumulation*, Oxford: Clarendon Press.

Bardhan, P.K. (1970) 'Trends in Land Reform', *Economic and Political Weekly*, 5, 261–6.

—— (1974) 'India', in Chenery, H., Ahluwalia, M., Bell, C., Duloy, J. and Jolly, R. (1974) *Redistribution with Growth*, Oxford: Oxford University Press.

—— (1985) 'Poverty and "Trickle Down" in Rural India: A Quantitative Analysis', in Mellor, J. and Desai, G. (eds) *Agricultural Change and Rural Poverty*, London: Johns Hopkins University Press.

—— (1974) 'India', in Chenery, H., Ahluwalia, M., Bell, B., Duloy, J. and Jolly, R. (eds) *Redistribution with Growth*, Oxford: Oxford University Press, Annex pp. 255–62.

Barraclough, S. (ed.) (1973) *Agrarian Structure in Latin America*, Lexington, KY: Lexington Books.

—— (1999) *Land Reform in Developing Countries: the Role of the State and Other Actors*, Geneva: UNRISD.

—— and Alfonso, A. (1972) *Critical Appraisal of the Chilean Agrarian Reform*, translated from Spanish, Santiago: Universidad Católica de Chile.

Barry, B. (2005) *Why Social Justice Matters*, Malden, MA: Polity Press.

Barry, N. (2002) 'The New Liberalism', in Smith, Gordon (ed.) *Liberalism*, Vol. II, London: Routledge.

Bates, R. (1981) *Markets and States in Tropical Africa: The Political Bases of Agricultural Politics*, Berkeley, CA: University of California Press.

Baumeister, E. (1994) *La Reforma Agraria en Nicaragua (1979–1989)*, Nijmegen: Katholieke Universiteit.

Baumol, W. (1965) *Welfare Economics and the Theory of the State*, London: Bell and Sons Ltd.

Berry, A. (ed.) (1998) *Poverty, Economic Reform and Income Distribution in Latin America*, London: Lynne Reinner Publishers.

—— and Cline, R. (1979) *Agrarian Structure and Productivity in Developing Countries*, Baltimore, MD: Johns Hopkins University Press.

Bhalla, S. (1979) 'Farm Size Productivity and Technical Change in Indian Agriculture', in Berry, A. and Cline, R. (eds) *Agrarian Structure and Productivity in Developing Countries*, Baltimore, MD: Johns Hopkins University Press, Appendix A, pp. 141–86.

Bigsten, A. (1983) *Income Distribution and Development*, London: Heinemann Educational Books.

Birdsall, N. and Londono, J. (1997) 'Asset Inequality Matters: An Assessment of the World Bank's Approach to Poverty Reduction', *The American Economic Review*, 87(2), 32–7.

Booth, Anne and Sandrum, P. (1985) *Labor Absorption in Agriculture*, New York: Oxford University Press.

Bromley, A. (1982) *Improving Irrigated Agriculture: Institutional Reform and Small Farmer*. World Bank Staff Working Paper 531, Washington, DC: World Bank.

—— (1989) 'The Other Land Reform', *World Development*, 17, 8.

Bruno, M., Ravallion, M. and Squire, L. (1996) *Equity and Growth in Developing Countries: Old and New Perspectives on the Policy Issues*. Policy Research Working Paper No. 1563, Washington, DC: World Bank.

Carter, M.R. (2000) 'Old Questions and New Realities: Land in Post-liberal Economics', in Zoomers, A. and van der Haar, G. (eds) *Current Land Policy in Latin America*, Amsterdam: KIT.

—— and Olinto, Pedro (1996) *Does Land Titling Activate a Productivity-promoting Land Market?* Working Paper, Madison, WI: Department of Agricultural Economics, University of Wisconsin.

—— and Salgado, Ramon (2002) 'Land Market Liberalization and the Agrarian Question in Latin America', in de Janvry, A., Gordillo, G., Platteau, J. and Sadoulet, E. (eds) *Access to Land Rural Poverty and Public Action*. UNU/WIDER Study, New York: Oxford University Press.

Cassen, R. and Associates (1986) *Does Aid Work?* Oxford: Clarendon Press.

Castillo, L. and Lehmann, D. (1983) 'Agrarian Reform and Structural Changes in Chile', in Ghose, A.K. (ed.) *Agrarian Reform in Contemporary Developing Countries*, London: Croom Helm.

Chebil, Mohsen 1967 'Evolution of Land Tenure in Tunisia in Relation to Agricultural Development Programs', in El-Ghonemy, Riad (ed.) *Land Policy in the Near East*, Rome: FAO.

Chen, S. and Wang, Y. (2001) *China Growth and Poverty Reduction: Trends between 1990 and 1999*. Policy Research Working Paper 2651, Washington, DC: The World Bank.

Chenery, H. (1979) *Structural Change and Development Policy*, Oxford: Oxford University Press.

——, Ahluwalia, Montek, Bell, Clive, Duloy, John and Jolly, Richard (1974) *Redistribution with Growth*, Oxford: Oxford University Press.

——, Ahluwalia, M., Duloy, J. and Jolly, R. (1981) *Redistribution with Growth*, fourth edition, Oxford: Oxford University Press.

Clark, Colin and Haswell, M. (1964) *The Economics of Subsistence Agriculture*, London: Macmillan.

Coase, R.H. (1960) 'The Problem of Social Cost', *Journal of Law and Economics*, 3, 1–44.

Collier, P. and Dollar, D. (2001) 'Can the World Cut Poverty in Half?' *World Development*, 29(11), 1787–1802.

—— and Lal, D. (1986) *Labour and Poverty in Kenya 1900–1980*, Oxford: Clarendon Press.

Commander, Simon (1987) *The State and Agricultural Development in Egypt since 1973*, London: Ithaca Press.

—— and Hadhoud, A.A. (1986) *Employment, the Labour Market and the Choice of Technology in Egyptian Architecture*, London: Overseas Development Institute.

Commons, John R. (1923) *Legal Foundations of Capitalism*, New York: Macmillan.

—— (1934) *Institutional Economics; its Place in Political Economy*, New York: Macmillan.

Cornia, Giovanni (1985) 'Farm Size, Land Yields and the Agricultural Production Function: An Analysis for Fifteen Developing Countries', *World Development*, 13(4), 513–34.

Cox, Terry (1986) *Peasants, Class and Capitalism*, Oxford: Clarendon Press.

Cross, C., Mingadi, Sibanda and Jama, V. (1996) 'Making a Living under Land Reform', in *Land, Labour and Livelihoods in South Africa*, Vol. II.

Currie, J.M. (1981) *The Economic Theory of Agricultural Land Tenure*, Cambridge: Cambridge University Press.

Dandekar, M. (1962) 'Economic Theory and Agrarian Reform', Oxford Economic Papers, 14, February.

—— (1964) *From Agrarian Reorganisation to Land Reform*, Vol. 6, 1, Artha Vijnama Poona, India: Gokhale Institute.

Dasgupta, P. (1987) 'Inequality as a Determinant of Malnutrition and Unemployment Policy', *The Economic Journal*, 97, 177–88.

—— (1995) *An Inquiry into Well-being and Destitution*, Oxford: Clarendon Press.

Datt, G., Joliffe, G. and Sharma, M. (2001) 'A Profile of Poverty in Egypt', *African Development Review*, 13(2), 202–37.

Deaton, A. (2001) 'Counting the World's Poor: Problems and Possible Solutions', *World Bank Research Observer*, 16(2).

Deere, C. *et al.* (1995) 'Household Incomes in Cuban Agriculture', *Development and Change*, 26, 209–34.

—— (1998) 'The Reluctant Reformer', in Szelenyi, I. (ed.) *Privatizing the Land*, London: Routledge.

Deininger, K. (1999) 'Making Negotiated Land Reform Work: Initial Experience from Colombia, Brazil and South Africa', *World Development*, 27(4), 651–72.

—— (2003) *Land Policies for Growth and Poverty Reduction*, Oxford: World Bank and Oxford University Press.

—— and Gonzalez, M. (2002) *Land Markets and Land Reform in Colombia*. Discussion paper, Washington, DC: World Bank.

—— and Olinto, P. (2002) chapter, in Deininger, K. (ed.) *Land Policies for Growth and Poverty Reduction*, World Bank and Oxford University Press.

—— and Squire, L. (1998) 'New Ways of Looking at Old Issues: Inequality and Growth', *Journal of Development Economics*, 57(2), 259–87.

—— and Zegarra, E. (2003) 'Determinants and Impacts of Rural Land Market in Nicaragua', in Deininger, K. (ed.) *Land Policies for Growth and Poverty Reduction*, World Bank and Oxford University Press.

Demsetz, H. (1982) 'Barriers to Entry', *The American Economic Review*, 72(1).

Denman, D. (1973) *The King's Vista – A Land Reform which has Changed the Face of Persia*, London: Geographical Publications Ltd.

De Waal, Alexander (1989) *Famine That Kills, Darfur, Sudan, 1984–1985*, Oxford: Clarendon Press.

Dong Wang S. and Yang Boo, C. (1984) *Alleviation of Rural Poverty in South Korea*. An FAO Rural Poverty Study, Rome: FAO.

Dorner, P. (1964) 'Land Tenure, Income Distribution and Productivity Interconnection', *Land Economics*, 40 (3).

—— and Kanel, D. (1971) 'The Economic Case of Land Reform', *Land Reform and Land Settlement*, 1, 59–68.

Dovring, E. (1970) *Land Reform in Mexico*. Agency for International Development (USAID), Spring Review, Washington, DC: USAID.

Dreze J. and Sen, A. (1995) *Economic Development and Social Opportunities in India*, Oxford: Clarendon Press.

DSE (Deutsche Stiftung für Internationale Entwicklung) (2001) 'Access to Land Provides Food Security', *Development Cooperation*, 4, 26.

Dumont, R. (1964) *Cuba: Socialism et Development*, Paris: Editions du Seuil.

Egypt's Central Statistical Organization (CAPMAS), Statistical Yearbook, several years, Cairo: CAPMAS.

El-Ghonemy, M. Riad (1953) Resource use and income in Egyptian agriculture before and after land reform. Unpublished PhD thesis, North Carolina State University, Raleigh, NC.

—— (ed.) (1967) *Land Policy in the Near East*, Rome: FAO.

—— (1968a) 'Economic and Institutional Organization of the Egyptian Agriculture', in Vatikiotis, P.J. (ed.) *Egypt since the Revolution*, London: Allen & Unwin.

—— (1968b) 'Land Reform and Economic Development in the Near East', *Land Economics* XLIV(1), 36–49.

—— (ed.) (1984a) *Development Strategies and the Rural Poor*. Development Paper No. 44, Rome: FAO.

—— (1984b) 'The Crisis of Rural Poverty: can Participation Resolve it?', in El-Ghonemy, M.R. (ed.) *Studies on Agrarian Reform and Rural Poverty*, Rome: FAO.

—— (ed.) (1986) *The Dynamics of Rural Poverty*, Rome: FAO.

—— (1990a) *The Political Economy of Rural Poverty: The Case for Land Reform*, London: Routledge.

—— (1990b) 'Egyptian land reform and its relevance to the Philippines', in *International Issues in Agrarian Reform: Past Experience and Future Prospects*, Laguna, Philippines: The Government of the Philippines, Institute of Agrarian Studies, University of Los Banos; and Rome: FAO.

—— (1992) 'The Egyptian State and Agricultural Land Market 1810–1986', *Journal of Agricultural Economics*, 43(2), 175–90.

—— (1993) *Land, Food, and Rural Development in North Africa*, Boulder, CO: Westview Press.

—— (1998) *Affluence and Poverty in the Middle East*, London: Routledge.

—— (1999) 'The Political Economy of Market-Based Land Reform', as an UNRISD Publication and in Ghimire, K. (ed.) (2001) *Land Reform and Peasant Livelihoods*, London: ITDG Publishing.

—— (2002) 'The Land Market Approach to Rural Development', a paper presented at the Agrarian Relations Conference, at the invitation of the Indian Statistical Institute,

Kolkata, January 2002, and in Ramachandran, V.K. and Swaminathan, M. (eds) *Agrarian Studies*, New Delhi: Tulika Books.

—— (2003) *Egypt in the Twenty-first Century: Development Challenges*, London: Routledge.

—— (2005) 'Agriculture and Rural Poverty in Syria', in *Macroeconomic Policies for Poverty Reduction: The Case of Syria*, New York: UNDP.

——, Tyler, G. and Couvreur, Y. (1993) 'Alleviating Rural Poverty through Agricultural Growth', *Journal of Development Studies*, 29.

El-Leithy, H. *et al.* (2002) *Poverty Reduction in Egypt*, World Bank Report No. 24254. A joint study by the Ministry of Planning of Egypt and the Social and Economic Development Group of the World Bank.

——, Lokshin, M. and Banerji, A. (2003) *Poverty and Economic Growth in Egypt*. Policy Research Working Paper 3068, Washington, DC: World Bank.

El-Zoobi, A. (1984) *Alleviation of Rural Poverty through Agrarian Reform, Syria*. FAO Series on Rural Poverty, Rome: FAO.

FAO (Food and Agriculture Organization of the United Nations) (1950) *Production Yearbook*, Vol. IV, Washington, DC: FAO.

—— (1951) *Production Yearbook*, Vol. V, Rome: FAO.

—— (1979) *Report. World Conference on Agrarian Reform and Rural Development (WCARRD)*, Rome: FAO.

—— (1980) *Agriculture Toward 2000*, Rome: FAO.

—— (1984) *Poverty Alleviation in Yemen*. Poverty Studies 9, Rome: FAO.

—— (1985a) *Fifth World Food Survey*, Rome: FAO.

—— (1985b) *Report of the WCARRD Follow-up Inter-Agency Mission to Mozambique*, prepared by R. El-Ghonemy, Rome: FAO.

—— (1985c) *Production Yearbook and Country Tables*, Rome: Statistics Division, FAO.

—— (1986) *Landlessness, Problems and Policies*. Report of FAO Expert Consultation, Rome: FAO.

—— (1987) *Country Tables*, Rome: Statistics Division, FAO.

—— (1988a) *Guidelines on Socio-Economic Indicators for Monitoring and Evaluating Agrarian Reform and Rural Poverty*, Rome: FAO.

—— (1988b) *Women in Food Production*, Rome: FAO.

—— (1991) *Country Tables: Basic Data*, Rome: FAO.

—— (1993a) *Country Tables: Basic Data on the Agricultural Sector*, Rome: FAO.

—— (1993b) *Agriculture: Towards 2010*. FAO Conference Publication, Rome: FAO.

—— (1996a) *Food, Agriculture and Food Security, the Balance Between Population and Food Production*. World Food Summit, 13–17 November, Rome: FAO.

—— (1996b) *The Sixth World Food Survey*, Rome: FAO.

—— (1997a) *Report of the 1990 World Census of Agriculture: International Comparison and Primary Trends Results by Country, 1986–1995*, Rome: FAO.

—— (1997b) *Production Yearbook*, Rome: FAO.

—— (1999) *Filling the Data Gap: Gender-Sensitive Statistics*, Rome: FAO.

—— (2001) *FAO Statistical Yearbook*, Rome: FAO.

—— (2003a) *World Agriculture Towards 2015/2030*. An FAO Perspective prepared by J. Bruinsma, FAO. London: Earthscan Publishing Ltd.

—— (2003b) *Production Yearbook, Vol. 57*, Rome: FAO.

—— (2004a) *FAO Statistical Yearbook*, Rome: FAO.

—— (2004b) *The State of Food Insecurity in the World*, Rome: FAO.

FAO and WHO (1992) *Nutrition and the Global Challenge and Nutrition and Develop-ment*. International Conference on Nutrition, 5–11 December, Rome: FAO.

—— (2004) 'Estimates of Undernutrition', in FAO, *The STate of Food Insecurity in the World*, Rome: FAO.

Fedder, G. and Nishio A. (1997) 'The Benefits of Land Titling and Registration', *Land Use Policy*, 15(1), 143–69.

—— and Noronha, R. (1987) 'Land Rights Systems and Agricultural Development in sub-Saharan Africa', *World Bank Research Observer*, 2(2), 143–69.

Fenelon, K. (1970) 'Iraq National Income and Expenditure', in *National Income of Iraq, Selected Studies*, Baghdad: Central Statistical Organization, Government of Iraq.

Fields, G. (1980) *Poverty, Inequality and Development*, Cambridge: Cambridge University Press.

Forni, N. (2003) 'Land Tenure and Labour Relations', in *Syrian Agriculture at the Cross-roads*, FAO Agricultural Policy Series 8, Rome: FAO.

Forsyth, M. (2002) 'Hayek's Bizarre Liberalism: a Critique', in Smith, Gordon W. (ed.) *Liberalism: Critical Concepts in Political Theory*, London: Routledge.

Friedman, M. (1962) *Capitalism and Freedom*, Chicago: University of Chicago Press.

Gabbay, R. (1978) *Communism and Agrarian Reform in Iraq*, London: Croom Helm.

Gadalla, S. (1982) *Land Reform in Relation to Social Development in Egypt*. University of Missouri Studies No. 5, Columbia: University of Missouri.

Galbraith, J.K. (1984) *The Anatomy of Power*, London: Hamish Hamilton.

Gayoso, A. (1970) *Land Reform in Cuba*, Washington, DC: USAID.

Georgescu-Rogen, N. (1960) 'Economic Theory and Agrarian Economics', *Oxford Eco-nomic Papers*, xii, 1–40.

Ghai, D. (ed.) (1979) *Agrarian Systems and Rural Development*, London: Macmillan.

——, Kay, C. and Peek, P. (1988) *Labour and Development in Rural Cuba*. An ILO study, London: Macmillan.

Ghimire, K. (ed.) (2001) *Land Reform and Peasant Livelihoods*, London: UNRISD and ITDG Publishing.

Ghose, K. (ed.) (1983) *Agrarian Reform in Contemporary Developing Countries*, London: Croom Helm.

Government of Mozambique (1981, 1982, 1983) *State Farms' Marketed Crops*, Maputo: Agricom.

Government of the Philippines (1990) *Agrarian Reform Programme: Salient Features and Progress Made*. Report of an International Colloquium, Quezon City: Department of Agrarian Reform.

Government of South Africa (1995) *The Composition and Persistence of Poverty in Rural South Africa: An Entitlement Approach to Poverty*. Policy Paper No. 15, Pretoria: Land and Agriculture Policy Centre.

—— (1996) *Our Land*. Green Paper on South Africa Land Policy, n.p.: Department of Land Affairs.

Gray, A. (1931) *The Development of Economic Doctrine*, London: Longmans, Green & Co.

Griffin, Keith (1976) 'Income Inequality and Land Redistribution in Morocco', in *Land Concentration and Rural Poverty*, London: Macmillan.

—— (ed.) (1984) *Institutional Reform and Economic Development in the Chinese Coun-tryside*, London: Macmillan.

—— (1986) 'Communal Land Tenure Systems and their Role in Rural Development', in Lal, D. and Stewart, F. (eds) *Theory and Reality in Development*, London: Macmillan.

—— and Ghose, A.K. (1979) 'Growth and Impoverishment in the Rural Areas of Asia', *World Development*, 7, 361–83.

Guardian, E.A. (2003) 'Impacts of Access to Land for Food Security and Poverty: The Case of Philippine Agrarian Reform', in *Land Reform*, 2, 71–82.

Guevara, E.C. (1964) 'The Cuban Economy: Its Past and Present Importance', *International Affairs*, 40(4), October.

Guiltiano, A. (2003) *Population Dynamics, Land Availability and Adapting Land Tenure System*, Rome: FAO Land Tenure Service.

Haider, S. (1944) 'The Problem of the Land in Iraq', unpublished PhD Thesis, University of London.

Hansen, B. and Marzouk, G. (1965) *Development and Economic Policy in the UAR (Egypt)*, Amsterdam: North-Holland Publishing Co.

Harriss-White, B. (2003) *India Working: Essays in Economy and Society*, Cambridge: Cambridge University Press.

Hasseeb, K. (1964) *The National Income of Iraq 1953–61*, Oxford: Oxford University Press.

Hayami, Y. and Kikuchi, M. (1985) 'Directions of Agricultural Change in the Philippines', in Mellor, J. and Desai, G. (eds) *Agricultural Change and Rural Poverty*, Baltimore, MD: Johns Hopkins University Press.

Hayek, F. von (1978) *The Mirage of Social Justice*, London: Routledge and Kegan Paul.

—— (1986) *The Road to Serfdom*, London: Ark.

Henry, C.M. (1986) *Economies of Scale and Agrarian Structure*. Oxford Agrarian Studies, Oxford: Oxford University Institute of Agricultural Economics.

Hopkins, N. (1987) *Agrarian Transformation in Egypt*, Boulder, CO: Westview Press.

Horowitz, A. (1993) 'Time Paths of Land Reform: A Theoretical Model of Reform Dynamics', *The American Economic Review*, 83(4), 1003–10.

Hymer, S. (1971) 'The Multinational Corporation and the Law of Uneven Development', in Bhagawati, J. (ed.) *Economics and World Order*, New York: Macmillan.

IFAD (International Fund for Agricultural Development) (1992) *The State of World Rural Poverty: an Inquiry into the Causes and Consequence*, Jazairy, I., Alamgir, M. and Panuccio, T. (eds), Rome: IFAD.

—— (2001) *Rural Poverty Report: The Challenge of Ending Rural Poverty*, Oxford: Oxford University Press.

IFAD/World Bank/FAO (1997) *Network Proposal on Negotiated Land Reform*, Rome: IFAD/World Bank/FAO.

ILO (International Labour Organization) (1971) *Agrarian Reform and Employment*, Geneva: ILO.

IMF (International Monetary Fund) *Government Finance Statistics*, several issues, Washington, DC: IMF.

INS (Institut National de Statistiques) (1982) *Socio-economic indicators*, Tunis: INS.

Institute of Pacific Relations (1939) *Agrarian China: Selected Source Materials from Chinese Authors*, London: George Allen and Unwin.

Iradian, Garbis (2005) *Inequality, Poverty and Growth: Cross Country Evidence*. IMF Working Paper 28, Washington, DC: IMF.

Islam, R. (ed.) (1985) *Strategies for Alleviating Poverty in Rural Asia*, Dhaka: ILO and Bangladesh Development Institute.

Issawi, Charles (1982) *An Economic History of the Middle East and North Africa*, New York: Colombia University Press.

de Janvry, A. (1981) *The Agrarian Question and Reformism in Latin America*, Baltimore, MD: Johns Hopkins University Press.

——, Gordillo, Gustavo, Platteau, J. and Sadoulet, Elizabeth (eds) (2002) *Access to Land, Rural Poverty and Public Action*, Oxford: Oxford University Press.

Johnson, C. (1987) 'Political Institutions and Economic Performance: A Comparative Analysis of the Government–Business Relationship in Japan, South Korea and Taiwan', in Deyo F. (ed.) *The Political Economy of the Asian Industrialization*, Ithaca: Cornell University Press.

Jose, A.V. (1983) 'Poverty and Inequality: The Case of Kerala', in Khan, A.R. and Lee, E. (eds) *Poverty in Rural Asia*, Geneva: ILO.

Joshi, P.C. (1975) *Land Reforms in India: Trends and Perspectives*, Delhi: Allied Publishers.

Kaldor, N. (1935) 'Market Imperfection and Excess Capacity', *Economica*, II, February.

Kalecki, M. (1934) *Theory of Economic Dynamics*, London: Unwin University Books.

Katouzian, M. (1983) 'The Agrarian Question in Iran', in Ghose, A. (ed.) *Agrarian Reform in Contemporary Developing Countries*, London: Croom Helm.

Kaufman, H.F. *et al.* (1953) 'Problems of Theory and Method in the Study of Social Stratification in Rural Society', *Rural Sociology*, 18.

Kay, C. (1983) 'The Agrarian Reform in Peru: an Assessment', in Ghose, A.K. (ed.) *Agrarian Reform in Contemporary Developing Countries, an ILO Study*, London: Croom Helm.

Keidel, A. (1981) *Korean Regional Farm Production and Income 1910–1975*, Seoul: Korea Development Institute.

Keynes, J.M. (1936) *The General Theory of Employment, Interest and Money*, New York: Harcourt Brace and Co.

Khan, A.R. (1984) *The Responsibility System and Economic Development in the Chinese Countryside*, Macmillan.

—— and Lee, E. (eds) (1983) *Poverty in Rural Asia*, Bangkok: ARTEP.

Kifle, H. (1983) 'The Role of State Farms in Agrarian Transformation in Centrally Planned Economies of Africa', a study prepared for a Regional Workshop organized by FAO in Arusha, Tanzania, 17–22 October 1983.

Kim, Wan-Soon and Yun, K.Y. (1988) 'Fiscal Policy and Development in Korea', *World Development*, 16(1), 65–83.

Kuznets, S. (1955) 'Economic Growth and Economic Inequality', *American Economic Review*, 65(1), 1–28.

Lampton, A.K. (1958) *Landlord and Peasant in Persia*, Oxford: Oxford University Press.

Land and Agriculture Policy Centre (1995) 'The Composition and Persistence of Poverty in Rural South Africa: An Entitlement Approach to Poverty'. Policy Paper 15, Pretoria: Government of South Africa.

Lardy, N.R. (1983) *Agriculture in China's Modern Economic Development*, Cambridge: Cambridge University Press.

Lazarev, G. (1977) 'Aspects du Capitalisme Agraire au Maroc avant le Protectorat', in Bruno, Etienne (ed.) *Les Problemes Agraires au Maghreb*, Paris: Centre National de la Recherche Scientifique.

Ledesma, A.S. (1982) *Landless Workers and Rice Farmers*, Los Banos, Philippines: International Institute for Rice Research.

Lee, E. (1979) 'Egalitarian Peasant Farming and Rural Development in South Korea', in Ghai, D. (ed.) *Agrarian Systems and Rural Development*, London: Macmillan.

Lee, Hoon (1936) *Land Utilization and Rural Economy in Korea*, New York: Greenwood Press.

Lehman, D. (1978) 'The Death of Land Reform: A Polemic', *World Development*, 6(3), 339–45.

Lenin, V. (1899) 'The Development of Capitalism in Russia', reprinted in *Lenin: Collected Works*, 1960, Vol. 3, Moscow: Foreign Languages Publishing House.

Levin, R. (2002) 'Land and Agrarian Reform in South Africa', in Ramachandran, V.K. and Swaminathan, M. (eds) *Agrarian Studies*, New Delhi: Tulika Books.

Lewis, W. Arthur (1955) *The Theory of Economic Growth*, London: George Allen and Unwin.

Lipton, M. (1974) 'Towards a Theory of Land Reform', in Lehman, D. (ed.) *Agrarian Reform and Agrarian Reformism*, London: Faber & Faber.

—— (1977) *Why Poor People Stay Poor*, London: Temple Smith.

—— (1985) *Land Assets and Rural Poverty*. World Bank Staff Working Paper No. 744, Washington, DC: World Bank.

Longhurst, R. (1983) 'Agricultural Production and Food Consumption: Some Neglected Linkages', *Food and Nutrition*, 9(2), 1–5.

Lowe, J.W. (1977) 'The International Finance Corporation and the Agribusiness Sector', *Finance and Development*, 26.

Lugogo, J.A. (1986) 'The Impact of Structural Changes in Kenya's Plantation Sector', in *The Socio-Economic Implications of Structural Changes in Plantations in African Countries*. Working Paper, Geneva: ILO.

McCormic, C. (1956) 'It can be Solved', in Parsons, K.H., Penn, R.J. and Raup, P.M. (eds) *Land Tenure – Proceedings of the International Conference on Land Tenure*, held at Madison, Wisconsin, USA.

McEntire, D. and McEntire, I.L. (1969) 'Agrarian Reform in Mexico', in *Toward Modern Land Policies*, Padua, Italy: University of Padua.

McGraham, D., Wolf, S. and Claude, R. (1985) *Qualitative Indicators of Development*. Discussion Paper No. 15, Geneva: United Nations Research Institute for Social Development (UNRISD).

MacLeod, H.D. (1867) *The Elements of Economics*, London.

Mangahas, M. (1985) 'Rural Poverty and Land Transfer in the Philippines', in Islam, R. (ed.) *Strategies for Alleviating Poverty in Rural Asia*, Bangkok: ARTEP.

—— and Barros, B. (1980) *The Distribution of Income and Wealth*, Manila: The Philippines Institute for Development Studies.

Marshall, A. (1952) *Principles of Economics*, 8th edn, London: Macmillan.

Marx, K. (1906) *Capital: A Critique of Political Economy*, New York: The Modern Library.

Matthews, R.C.O. (1986) 'The Economics of Institutions and the Sources of Growth', *The Economic Journal*, 96, 903–18.

Mellor, J. (1975) *Agricultural Price Policy and Income Distribution in Low Income Nations*. World Bank Staff Working Paper No. 214, Washington, DC: World Bank.

—— and Desai, G. (eds) (1985) *Agricultural Change and Rural Poverty*, Baltimore, MD: Johns Hopkins University Press.

—— and Johnson, B. (1984) 'The World Food Equation: Interrelations among Development, Employment and Food Consumption', *Journal of Economic Literature*, 22.

Melville, B. (1988) 'Are Land Availability and Cropping Pattern Critical Factors in Determining Nutritional Standards', *Food and Nutrition Bulletin*, Tokyo: The United Nations University.

Mill, J.S. (1948) *Principles of Political Economy*, Penguin Books 1970 (Books IV and V), reprinted from Routledge and Kegan Paul (1965).

Ministry of Agrarian Reform of the Philippines (1982) *Agrarian Reform Programme: Highlights of Accomplishments*, Manila: Ministry of Agrarian Reform.

Modigliani, F. (1975) *Determinants of National Savings and Wealth*, London: Macmillan.

—— (1988) 'Measuring the Contribution of Intergenerational Transfers to Total Wealth', in Kessler, D. and Masson, A. (eds) *Modelling the Accumulation of Wealth and Distribution*, Oxford: Oxford University Press.

Moghadam, E. (1978) 'The Effect of Farm Size and Management System on Agricultural Production in Iran', unpublished PhD thesis, University of Oxford.

Montgomery, J.D. (ed.) (1984) *International Dimensions of Land Reform*, Boulder, CO: Westview Press.

Morocco Ministry of Agriculture (1973/1974) *Recensement Agricole*, Rabat: Morocco Ministry of Agriculture.

—— (1987) *L'Agriculture en Chiffres*, Rabat: Morocco Ministry of Agriculture.

—— (1989) *al-felaha fi tanmiyya mostamerah (Agriculture in Continual Development)*, Rabat: Morocco Ministry of Agriculture.

—— (1990) *Rapport sur l'avancement du developpement rural*, Rabat: Morocco Ministry of Agriculture.

—— (1984–1985) *Consommation et Depenses des Menages Premiers Resultats*, Vol. 1, Rapport de synthese, Rabat: Morocco Ministry of Agriculture.

Morocco Ministry of Planning (1960–1964) *First Development Plan*, Rabat: Morocco Ministry of Planning.

Morowitz, D. (1977) *Twenty-Five Years of Economic Development, 1950–1975*, Washington, DC: World Bank.

Mosley, P. (1999) 'Micro–Macro Linkages in Financial Markets: The Impact of Financial Liberalization on Access to Rural Credit in four African Countries', *Journal of International Development*, 11(3), 367–84.

Murray, J. and Lopez, A. (eds) (1996) *The Global Burden of Disease*, Cambridge, MA: Harvard University Press for WHO, Harvard School of Public Health and World Bank.

Myers, R. (1982) 'Land Property Rights in Modern China', in Barker, R. *et al.* (eds) *The Chinese Agricultural Economy*, London: Croom Helm.

Myrdal, G. (1956) *An International Economy, Problems and Prospects*, London: Routledge and Kegan Paul.

—— (1960) *Beyond the Welfare State*, New Haven, CT: Yale University Press.

—— (1968) *Asian Drama, An Enquiry into the Poverty of Nations*, New York: Twentieth Century Fund.

Najafi, B. (1970) 'The Effectiveness of Farm Corporations in Iran', unpublished MSc thesis, American University, Beirut.

Naschold, F. (2004) 'Growth Distribution and Poverty Reduction: LDCs are Falling Further Behind', in Shorrocks, A. and van der Hoeven, R. (eds) *Growth, Inequality and Poverty*, Oxford: Oxford University Press.

Nickolsky, S. (1998) 'The Treadmill of Socialist Reforms and the Failures of Post-Communist Revolutions in Russian Agriculture', in Szelenyi, I. (ed.) *Privatizing the Land*, London: Routledge.

North Korean Yearbook of Agriculture (1980) Pyongyang: Ministry of Agriculture.

Nouvele, J. (1949) 'La Crise Agricole de 1843–1946 au Maroc et ses Consequences Economiques et Sociales', *Revue de Geographie Humaine*, 1, 87–9.

Nozick, R. (1974) *Anarchy, State and Utopia*, Oxford: Basil Blackwell.

Nsabagasani, X. (1997) *Land Privatization, Security of Tenure and Agricultural Production: The Ugandan Experience*, The Hague: Institute of Social Studies.

Oakley, P. and Marsden, D. (1984) *Approaches to Participation in Rural Development*, Geneva: ILO.

OECD/DAC (2000) *International Statistics*, OECD.

Olson, G.L. (1974) *US Foreign Policy and the Third World Peasant*, Westport, CT: Praeger.

Orr, A., Mwale, B. and Saiti, B. (2001) 'Market Liberalization, Household Food Security and the Rural Poor in Malawi', *The European Journal of Development Research*, 13(1), 47–69.

Palmer, I. (1983) *Rainfed Agricultural Development in Thailand: A Survey of Women's Role*, New York: Population Council.

Parsons, K. (1962) *Agrarian Reform and Economic Growth in Developing Countries*, Washington, DC: US Department of Agriculture.

Parsons, L.H., Penn, R. and Raup, P. (eds) (1956) *Land Tenure: Proceedings of the International Conference on Land Tenure and Related Problems*, Madison, WI: The University of Wisconsin Press.

—— (1965) 'The Tunisian Program for Co-operative Farming', *Land Economics*, XLI.

—— (1984) 'The Place of Agrarian Reform in Rural Development Policies', in El-Ghonemy, M.R. (ed.) *Studies on Agrarian Reform and Rural Poverty*, Rome: FAO.

—— (1985) 'John R. Commons: his Relevance to Contemporary Economics', *Journal of Economic Issues*, September.

Pascon, Paul (1977) *La Houaz de Marrakesh*, Rabat.

—— and Ennaji, M. (1986) *Les Paysans sans Terre au Maroc: Connaisance Sociale*, Rabat: Les Editions Tubkal.

Peek, P. (1984) *Collectivising the Peasantry*, Geneva: ILO.

Perkins, D. and Yusuf, S. (1984) *Rural Development in China*. A World Bank Publication, Baltimore, MD: Johns Hopkins University Press.

Perkins, K. (1989) *Historical Dictionary of Tunisia*, London: The Scarecrow Press.

Perotti, R. (1992) 'Income Distribution, Politics and Growth', *American Economic Review Papers and Proceedings*, 82, 311–16.

Phelps Brown, H. (1988) *Egalitarianism and the Generation of Inequality*, Oxford: Clarendon Press.

Place, F. and Hazell, P.B. (1993) 'Productivity Effects of Indigenous Land Tenure Systems in Sub-Saharan Africa', *American Journal of Agricultural Economics*, 75, 10–19.

—— and Otsuka, K. (2000) 'Population Pressure, Land Tenure, and Tree Resource Management in Uganda', *Land Economics*, 76, 233–51.

—— and Otsuka, K. (2001) Tenure, Agricultural Investment, and Productivity in the Customary Tenure Sector of Malawi', *Economic Development and Cultural Change*, 50, 77–100.

Platteau, J. (1995) *Reforming Land Rights in Sub-Saharan Africa: Issues on Efficiency and Equity*. Discussion Paper No. 60, Geneva: UNRISD.

—— (1996) 'The Evolutionary Theory of Land Rights as Applied to Sub-Saharan Africa', *Development and Change*, 27(1).

Putzel, James (2002) 'The Politics of Partial Land Reform in the Philippines', in Ramachandran, V.K. and Swaminathan, M. (eds) *Agrarian Studies*, New Delhi: Tulika Books.

Radwan, Samir (1969) *Agrarian Reform and Rural Poverty: Egypt, 1952–1975*. An ILO study, Geneva: ILO.

—— (1977) *Agrarian Reform and Rural Poverty: Egypt, 1952–1975*, Geneva: ILO.

—— and Lee, Eddy (1986) *Agrarian Change in Egypt: An Anatomy of Rural Poverty*, London: Croom Helm.

—— and Thomson, Anne (1981) *Aide Memoire on the Food Subsidy System in Morocco*. An IDS and ILO joint study, Geneva: ILO.

——, Jamal, Vali and Ghose, Ajit (1991) *Tunisia: Rural Labour and Structural Transformation*, London: Routledge.

Rahman, Sobhan (1983) *Rural Poverty and Agrarian Reform in the Philippines*. FAO Rural Poverty Series, Rome: FAO.

Ramachandran, V.K. and Swaminathan, M. (eds) (2002) *Agrarian Studies*, New Delhi: Tulika Books.

Ravallian, M. (1987) *Markets and Famine*, Oxford: Clarendon Press.

Rawls, J. (1972) *The Theory of Justice*, Oxford: Oxford University Press.

Ritter, A. (1980) *Anarchism: A Theoretical Analysis*, Cambridge: Cambridge University Press.

Roemer, John (1982) *A General Theory of Exploitation and Class*, Cambridge, MA: Harvard University Press.

—— (1996) *Theories of Distributive Justice*, Cambridge, MA: Harvard University Press.

Roth, M.J. (2003) 'Finding Solutions and Securing Rights in South Africa's Land Reform', *Land Tenure Newsletter*, University of Wisconsin, Madison.

Russell, B. (1938) *Power: A New Social Analysis*, New York: Norton.

Ruttan, Vernon (1975) 'Integrated Rural Development Programs: A Historical Perspective', *World Development*, 12, 393–401.

Saab, Gabriel (1976) *The Egyptian Agrarian Reform, 1952–62*, Oxford: Oxford University Press.

Safilios-Rothschild, C. (1983) *Women in Agrarian Reform in Honduras*. Land Reform and Settlement Bulletin 1/2, Rome: FAO.

Saith, A. (1984) 'China's New Population Policy', in Griffin, K. (ed.) *Institutional Reform and Economic Development in China's Countryside*, London: Macmillan.

—— (1989) *Development Strategies and the Rural Poor*. Working Paper 66, The Hague: Institute of Social Studies.

Scheftel, E. (1947) *The Jewish Law of Family and Inheritance*, Tel Aviv.

Schultz, T.W. (1964) *Transforming Traditional Agriculture*, New Haven, CT: Yale University Press.

Sen, A.K. (1966) 'Size of Holding and Productivity', *Economic Weekly*, 2.

—— (1981) *Poverty and Famine: An Essay on Entitlement Deprivation*, Oxford: Clarendon Press.

—— (1982) *Choice, Welfare and Measurement*, Oxford: Basil Blackwell.

—— (1983) 'Development Which Way Now', *The Economic Journal*, 2(72), 745–62.

—— (1984) *Resources, Values and Development*, Oxford: Oxford University Press.

—— (2000) 'Social Justice and the Distribution of Income', in Atkinson, A. and Bourguignon, F. (eds) *Handbook of Income Distribution*, Vol. 1, Amsterdam: North Holland.

—— (2002) 'Agriculture, Employment and Poverty: Recent Trends in Rural India', in Ramachandran, V.K. and Swaminathan, M. (eds) *Agrarian Studies*, New Delhi: Tulika Books.

—— and Anand, S. (2000) 'Human Development and Economic Sustainability', *World Development*, 28(12), 2029–49.

—— and Nussbaum, M. (eds) (1993) *The Quality of Life*, Oxford: Clarendon Press.

Sentra (Centro Pata SA Tunay na Reportamg Agraryo) (1997) *Market-Oriented CARP in Philippines: A Recipe for Failure*, Manila: Sentra.

Shirazi, A. (1993) *Islamic Development Policy: The Agrarian Question in Iran*, Boulder, CO: Lynne Rienner.

Shorrocks, A. and van der Hoeven, R. (2004) *Growth, Inequality and Poverty: Prospects for Pro-poor Economic Development.* A study prepared by UNU/WIDER, Oxford: Oxford University Press.

Singer, H. (1992) Lessons of Post-War Development Experience till 1988', in Andriessen, W. (ed.) *A Dual World Economy: Forty Years of Development Experience*, Bombay: Oxford University Press.

Sinha, R. (1984) *Landlessness: A Growing Problem*, Rome: FAO.

Sjaastad, E. (2003) 'Trends in the Emergence of Agricultural Land Markets in Sub-Saharan Africa', *Forum of Development Studies*, 1(1), 5–27.

Smith, Adam (1776) *The Wealth of Nations*, n.p.: Random House (1937).

Smith, Gordon (ed.) (2002) *Liberalism*, London: Routledge.

Sobhan, R. (ed.) (1988) *Labour and Labour Unions in Asia and Africa: Contemporary Issues*, London: Macmillan.

SOFA (1970 and 1998) *The State of Food and Agriculture*, Rome: FAO.

Solow, Robert (1991) 'Sustainability: An Economist's Perspective', The 18th Steward Jonson Lecture, Woods Hole, MA, USA.

Srinivasan, T. (2000) 'Growth, Poverty Reduction and Inequality', Department of Economics, Yale University, USA (mimeographed).

Stiglitz, J. (2001) 'Redefining the Role of the State', in Chang, Ha-Joon (ed.) *Joseph Stiglitz and the World Bank, the Rebel Within*, London: Athen Press.

Streeten, P. (1972) *The Frontiers of Development Studies*, London: Macmillan.

Swearingen, W. (1988) *Moroccan Mirages: Agrarian Dreams and Development 1912–1986*, London: Tauris & Co.

Syrian Government (2001) *Rapport Syrien Economique*, Damascus: Syrian Government.

—— (2003) *Statistical Abstract*, Damascus: Syrian Government.

Szelenyi, I. (1998) *Privatizing the Land: Rural Political Economy in Post-communist Societies*, London: Routledge.

Taniguchi, K. (2003) *Nutrition Intake and Economic Growth. Studies on the Cost of Hunger*, Rome: FAO.

Tantawy, Mohamad Sayed (1992) *al-halal wal-haram fi mo' amalat al-bonouk*, Cairo: al-Ahram al-Iqtisadi.

Taussig, M. (1982) 'Peasant Economies and the Development of Capitalist Agriculture in Colombia', in Harriss, J. (ed.) *Rural Development and Agrarian Change*, London: Hutchinson University Library.

Thiesenhuesen, W.C. (1995) *The Broken Promises: Agrarian Reform and the Latin American Campesino*, Boulder, CO: Westview Press.

Tunisia, Ministry of Agriculture, *Annuaire des Statistiques Agricoles*, various issues. *Enquete Agricole de Base*, 1976 and 1980. Tunis: Ministere de l'Agriculture. 'Socio-Economic Indicators for Monitoring Agrarian Reform and Rural Development'. Institut National de Statistiques (mimeographed in Arabic).

Tyler, G.J. (1983) *Case Study on Rural Poverty in Somalia*, Rome: FAO.

Tyler, G., El-Ghonemy, M. Riad and Couvreux, Y. (1993) 'Alleviating Rural Poverty through Agricultural Growth', *The Journal of Development Studies*, 29, 359–63.

UNDP (1991 and 1992) *Human Development Report*, New York: Oxford University Press.

—— (2003) *Human Development Report*, New York: Oxford University Press.

218   *Bibliography*

UNECOSOC (Economic and Social Council of the UN) (1999) *Improvement of the Situation of Women in Rural Areas*. Report A154/123E, New York: UNECOSOC.
UNESCO (1998) *Statistical Yearbook*, Paris: UNESCO.
UNICEF (1984) *Statistics on Children in UNICEF Countries*, New York: UNICEF.
—— (1998) *The State of World Children*, UNICEF.
United Nations (1980) *Patterns of Urban and Rural Population Growth*, New York: UN.
—— *National Account Statistics*, several issues, New York: UN.
—— (2000) 'Redistributive Land Reform', in *World Economic and Social Survey*, New York: UN Department of Economic and Social Affairs.
—— (2003a) *Millennium Development Goals*, New York: UN.
—— (2003b) *Millennium Indicators. Database*, New York: UN Statistics Division.
—— (2005) *World Economic and Social Survey*, New York: UN Department of Economic and Social Affairs.
United Nations Fund for Population Activities (UNFPA) (1979) *Survey of Laws on Fertility Control*, New York: UNFPA.
UNRISD (UN Research Institute for Social Development) (1997) *Report of the International Network on Land Market Reform*, Geneva: UNRISD.
USAID (United States Agency for International Development) (1970) *Spring Review of Land Reform*, Washington, DC: USAID.
USAID (1986) *Policy Determination of the Agency for International Development: Land Tenure*. No. PD-13, Washington, DC: USAID.
Usher, D. (1981) *The Economic Prerequisites to Democracy*, Oxford: Blackwell.
da Veiga, J.E. (2003) 'Poverty Alleviation through Access to Land: The Experience of the Brazilian Agrarian Reform Process', *Land Reform and Land Settlement*, 2, 59–68.
Visaria, P. (1981) *Size of Landholding, Living Standard and Employment in Rural Western India*. World Bank Staff Working Paper No. 459, Washington, DC: World Bank.
Voelkner, H. and French, J. (1970) *A Dynamic Model for Land Reform Analysis*. Spring Review on Land Reform Paper 7, Washington, DC: USAID.
Vogelgesang, F. (1996) *Property Rights and the Rural Land Market in Latin America*. CEPAL Review No. 58, Santiago, Chile: CEPAL.
Von Braun, J. and de Haen, H. (1983) *The Effects of Food Price and Subsidy Policies in Egypt*. Research Report No. 42, Washington, DC: IFPRI.
Wade, R. (1990) *Governing the Market: Economic Theory and the Role of Government in East Asian Industrialization*, Princeton, NJ: Princeton University Press.
Waldron, Jeremy (1988) *The Right to Private Property*, Oxford: Clarendon Press.
Walker, K. (1965) *Planning in Chinese Agriculture, 1956–62*, London: Frank Cass.
Wan Dong, S. and Yang-Boo, C. (1984) *Alleviation of Rural Poverty in the Republic of Korea*. A study on poverty prepared for FAO, No. 12, Rome: FAO.
Warriner, Doreen (1948) *Land and Poverty in the Middle East*, London: Royal Institute of International Affairs.
—— (1969) *Land Reform in Principle and Practice*, Oxford: Clarendon Press.
White, Gordon (ed.) (1983) *Chinese Development Strategy after Mao in Revolutionary Socialist Development*, Brighton: Wheatsheaf.
White, H. and Black, R. (eds) (2003) *Targeting Development: Critical Perspective on the Millennium Development Goals*, London: Routledge.
Wolford, Wendy (2002) 'Case Study: Grass Roots Initiated Land Reform in Brazil', in de Janvry, A. Cordillo, A., Platteau, G. and Sadoulet, E. (eds) *Access to Land, Rural Poverty and Public Action*, New York: Oxford University Press.
World Bank (1975) *Land Reform Policy*, Washington, DC: World Bank.

—— (1980) *Aspects of Poverty in the Philippines: A Review and Assessment*, vol. 2, Washington, DC: World Bank.

—— (1983) *Focus on Poverty*, Washington, DC: World Bank.

—— (1986) *Poverty and Hunger: Issues and Options for Food, Security in Developing Countries*. A World Bank Policy Study, Washington, DC: World Bank.

—— (1986 and 1987) *Annual Report*, Washington, DC: World Bank.

—— (1987) *The World Bank Research Observer*, 2(2), 143–69.

—— (1988) *Rural Development: World Bank Experience, 1965–1986. A World Bank Operations Evaluation Study*, Washington, DC: World Bank.

—— (1992) *World Development Report*, Oxford: Oxford University Press.

—— (1993) *The East Asian Miracle: Economic Growth and Public Policy*, New York: Oxford University Press.

—— (1994) *Poverty in Colombia*, Washington DC: World Bank.

—— (1995) *From Scarcity to Security: Averting a Water Crisis*, Washington, DC: World Bank.

—— (1996) *Land Quality Indicators*. Discussion Paper No. 315, with FAO, UNDP and UNEP, Washington, DC: World Bank.

—— *World Development Report*, several issues, Oxford: Oxford University Press.

—— (2000) *World Development Report 2000/2001: Attacking Poverty*, New York: Oxford University Press.

—— (2001) *World Development Report: Attacking Poverty*, New York: Oxford University Press.

—— (2002) *World Development Report: Building Institutions for Markets*, New York: Oxford University Press.

—— (2005) *World Development Indicators*, Washington, DC: World Bank.

—— (2006) *World Development Report: Equity and Development*, New York: Oxford University Press.

Yunus, Mohammad (1994) *Credit is a Human Right*, Dhaka: Grameen Bank.

Zimmerman, F. (2000) 'Barriers to Participation of the Poor in South Africa Land Redistribution', *World Development*, 28(8), 1439–60.

Subject index

Afghanistan 184
Africa: North 3–4, 208, 211; South 12; sub-Saharan 3–4, 12, 209, 215
agrarian reform 9, 12, 129, 138, 161, 163–4, 194–6, 206; definition of 26–7
agricultural growth and poverty level 60–1, 103, 169, 208, 216–17
Albania 49
Algeria 52
Argentina 184, 190
Asia 4, 211; East 11; South 15
auctioneering land lease and sale 21, 120, 190

Bangladesh 3, 6, 8, 15, 45, 56, 176, 190, 194, 204
Benin 8, 183
betterment, rural: definition of 34–6, 182
birth control *see* population control
Bolivia: land (agrarian) reform 30, 80, 176–7, 188–9
Botswana 183
Brazil: efficiency by farm size 72; land concentration 57; land redistribution by market-based land reform 30, 151, 153, 167, 171, 217; by NGOs' assistance 181–2, 203, 218
budgetary cuts 179
Burundi 183
bureaucracy 134, 139, 165, 176, 203

Cameroon 186
Canada 64, 66, 149
capitalism 17, 23, 39, 63, 158, 182, 191, 194–7, 207, 213
central planning 18, 30, 111, 168–9
Chad 183

Chile 72, 176
China: complete land reform 81–90; food production 48, 86; land concentration 60; market liberalization 144, 148, 158–9, 166, 189; prevalence of rural poverty 3–4, 13; rural development path 130, 195
climate change effects 44–6, 48, 171
Colombia: arable land supply 52; land concentration 60; market-based land reform 152, 167, 207; rural poverty prevalence 11
collective farms 28, 110–12, 116, 176
command over food 33, 44, 177; resulting productivity of labour 33, 35, 39, 44, 177
communal land tenure 2, 100, 112–13, 175, 210
communist agrarian institutions 166
Congo 11
cooperatives in land reform areas 102, 110–11, 116, 122, 214
corruption 176
Costa Rica 184
credit supply (agricultural) 1, 18, 20, 42, 106, 120, 137–8
credit market 7, 58, 148, 161–3, 175
Cuba: complete land reform 110–12; food productivity 49; ideological preference 19, 146; land distribution 60; state farms 67–9, 193

demand for land 46, 51
deprivation 2
destitution 46
donor countries 144–5, 201
drought 44, 108, 111, 116, 134, 141 201

economic freedom 18, 177
economic growth and poverty reduction
 15, 214
economic theory related to land reform:
 institutions 19–20; Keynesian 40,
 192, 212; neoclassical 9, 17, 213;
 neoliberalism 17, 30–2, 177, 209, 216;
 welfare 38–9, 178, 205
economics of scale 71, 74–5
Ecuador 72, 184
education, access of the poor to 4, 66, 80
 110, 115, 182
efficiency, definition of 64; large farms
 50, 56, 63–5, 71, 103; small farms 48,
 63, 73
Egypt: arable land supply 56; land prices
 161; market orientation 114, 159–60,
 176–7; partial land reform 65, 83, 119–
 27; pre-land reform agrarian conditions
 30, 73, 119–27, 202; prevalence of
 rural poverty 6, 184, 197–8
entitlements 24
Ethiopia 44, 52, 183, 194
eviction of tenants 23, 129, 159
exploitation 17, 24, 110, 140; definition of
 23, 26–7; *see also* monopoly
expropriation of land 102, 110, 124, 133,
 140

family farm 102, 110, 140
family labour 66, 72, 124
famine 4, 45–6, 114, 135, 215
FAO (Food and Agriculture Organization
 of the UN) 7, 9, 12, 48, 54–5, 59, 61,
 69, 99, 109, 112, 146, 150, 164–6,
 193–4, 201, 204
farm size 63–6, 136, 207; related to food
 production 48, 50, 63, 70–6; size
 ceiling by land reform 80, 112, 115,
 121, 188, 198; *see also* efficiency
feudalism 23, 112, 194
Finland 190
food: aid *see* international aid; demand for
 40–1, 47, 76; deficit of 14; minimum
 nutritional requirements 3, 7, 44, 56;
 production per person 33, 48, 75–6,
 112, 137–8; subsidy 40, 129, 137, 198,
 219
France 51, 149; French rule 114, 133, 135
freedom of farm enterpreneurs 17–19, 177

Germany 149, 154, 193
Gini index of inequality (landholding/

income/expenditure): definition
 of 35–7, 55–7, 194; China 91,
 158–9; countries having high and low
 inequality 60, 194–5; Cuba 111; Egypt
 122, 127, 160, 202; Iraq 105, 109;
 Morocco 137; Russia 137; South Korea
 92, 95, 97
globalization (economic) 4, 171, 175, 179
government expenditures 28, 35, 50, 54,
 173
grabbing land 23, 112, 134
Guatemala 11
Gulf War, effects on rural labour market
 173
Guyana 74

Honduras: arable land supply 57; impact
 of market liberalization 144, 163;
 land concentration 163; rural poverty
 incidence 6, 184
human capabilities 76, 141, 182
human capital, rural areas 44, 141, 175,
 192
Hungary 11, 19, 193
hunger 8, 12, 46, 67, 134, 180, 186, 218

ideological preference 35, 110, 142,
 144–5, 176–7
IFAD (International Fund for Agricultural
 Development) 1, 4–5, 8, 154, 190, 201,
 203, 211
illiteracy 10, 38, 86, 131, 141, 182
ILO (International Labour Organization of
 UN) 53, 111, 117, 126, 203, 211
IMF (International Monetary Fund) 11,
 145, 149, 211
indebtedness 17, 58, 114, 120
India: partial land reform 129–31, 176,
 194; population growth control 171;
 rural development programme 198–9
Indonesia 6, 56
inflationary pressure 44, 120
inheritance 16, 23–6, 114, 141, 204
institutional constraints 55, 76
institutional monopoly 16, 19–21, 23, 42,
 58, 108, 177, 179
international aid 2, 11, 14–15, 78, 179,
 182, 190
Iran: partial land reform 80, 119, 200;
 estimate of rural poverty 184; tenancy
 132
Iraq: complete land reform 79–80, 99, 109,
 171, 176, 196, 205; poverty incidence

109–10, 184; tribal land 30
irrigation expansion 44, 50, 54, 102, 110, 112, 130, 135–6, 142; *see also* technological change
Italy 149
Ivory Coast 190

Jamaica 60
Japan 149, 180, 199, 202, 211
Jordan 6, 184, 190
judicial system 164
justice (social): land reform aim 37, 42, 142; theory of 37, 215

Kenya: arable land supply 52–4; labour employment and poverty 207; land concentration 56; large farmers' influence 73–7; rural poverty incidence 6, 183
Kuwait 173

labour market (agricultural) 20, 26, 46, 72–3, 148, 173, 177, 192
land concentration 2, 29, 55, 58, 108, 120, 137–8, 160–1; and agricultural growth 59–60; and poverty level 60, 169–71
land market 2, 18, 26, 30, 42, 161, 164, 181
land reform: definition of 26; complete and partial land reform, definition of 78–9, 118; voluntary land-market reform, definition of 30–2; assumptions used 174–7
land taxation 114, 161
land tenure, definition of 17; defective forms 20, 115, 153, 214
landless agricultural workers: definition of 5; intended land reform beneficiaries 80–1, 93, 116, 130; incidence of poverty and hunger 1, 4, 9, 33–4; estimated numbers 5, 6, 39, 44, 137–9, 216; likelihood of land purchase 161; wages and productivity 21, 24
Latin America 11, 13–14, 27, 55, 59, 75, 144, 197
Libya 48, 132
life expectancy 9–10, 80, 86, 131, 141, 182; *see also* human capabilities
Lorenz curve 37, 97, 109, 192, 194; *see also* Gini index of inequality

malnutrition/malnourishment 1, 45, 109, 150; *see also* nutrition

Mauritania 54
Mauritius 190
Mexico: food production 113; communal land (*ejido*) 112, 197; complete land reform, objectives 177; contents 79, 82, 112–13, 208, 213; poverty prevalence 184
Middle East 3, 13, 51, 173, 196, 211
migration/outmigration 50, 109, 158, 171, 195
Morocco: arable land supply 52; farm size productivity 72, 210; partial land reform 83, 133–7, 213; poverty incidence 136, 186; rural development 213
Mozambique 67, 70, 193
multinational corporations in agriculture: monopoly advantages 21–2; in Honduras 163; in Kenya 76–7; in Mexico 113; in the Philippines 139, 164; in South America 145; technical change 66

neoliberal economic policy 144
Nepal 56
Netherlands 190
Nicaragua 144, 157, 163, 205
Niger 44
Nigeria 176, 184
nomads 4, 46, 114, 182
non-governmental organizations 1, 138, 176, 187, 200–3
Norway 190
nutritional requirements 6–7, 33, 44, 108, 190; *see also* undernutrition

oil revenues' effect on rural workers' migration 100–3, 109, 132, 154
opportunity cost 64, 130

Pakistan: arable land supply 56, 71; irrigation expansion 172; land concentration 172; partial land reform 83, 172–3, 176; poverty estimates 3, 6, 184; relation to *Zakat* 174; role of tenants union in rent control 173
Panama 80
Paraguay 55, 60
pastoralists 4, 46, 115; *see also* nomads
population growth: demand for food and land 47, 171; pressure on land 56–8, 120, 161, 200; policy effects on poverty levels 2–5, 14; numbers of the rural

poor 2–3, 13; pace of reduction 131, 140, 171–2; prevalence *see individual countries*
privatization of collective farming 165, 202
privatization of communally held land 144, 155–6, 175, 178, 193, 214
privatization of state-owned farms 163, 165; *see also* neoliberalism

quality of life 106
Qur'an (Islamic principles): on inheritance 25–6

remittances of migrant farm workers 35, 46, 142, 173; in Egypt 160, 173, 198; in Pakistan 173
rental values paid by tenants 32, 120, 122, 129, 140, 159–60, 202
Romania 11, 193
rural development: agrarian defaults effects 136–9, 141–2; definition 34–6; non-farm employment component 35, 80, 124, 142, 173–4; poverty-reducing strategy 2, 26, 182; share in total expenditure 10, 148
Russia: ideological preference 19, 148; Lenin and agrarian communist institutions 23–4, 67; market orientation 144, 165–6

Saudi Arabia 57
scale of landownership redistribution 80, 82–3, 175
scarcity: of agricultural capital 75; of arable land 63, 66; of water for farming 1, 21, 123, 131
Sen's index of poverty: definition of 34; in Egypt 128; in India 130
sharecropping (sharecroppers) 29, 34, 55, 122, 132, 140–1, 198
socialism/socialists and agriculture: criteria for farm size efficiency 17, 63, 67–9, 102, 110; *see also* China; communist agrarian institutions; Cuba; Mozambique; Russia; state farms; Tunisia
Somalia 4, 185, 217
South Africa: arable land supply 53; market-based land reform 153–4, 171, 201, 210, 215; poverty estimate 151
South Korea: complete land reform 79–80, 91–8; and its complementarity with the market 180, 202; food productivity

48–9; poverty estimates 5, 98, 184; relationship with land inequality and agricultural GDP growth 92, 95; rural development 195
standardized market approach to land reform 144, 163; *see also* voluntary market-based land reform
state farms 67–71, 102–3, 116, 140, 176, 193; *see also individual countries*
starvation 44, 135; *see also* famine; undernutrition
Sri Lanka 15, 56, 176, 184, 190
Sudan: arable land supply 53; estimated poverty incidence 185; communally held land 155–6; famine 44, 53–4; farm size and productivity 194; international aid 11; nomads and pastoralists 4–5; water rights in agriculture 191
Swaziland 8
Sweden 190
Syria: estimate of landless agricultural workers 141; irrigated land expansion 142; partial land reform 119, 140–2; poverty estimate 141–2

Tanzania 183, 194, 212
technological change in reformed areas 54, 74–5; labour-saving 61, 66, 173, 179; yield-increasing 17, 172, 195; *see also* irrigation expansion
tenancy 27, 29, 38, 100, 114, 129, 141, 175, 178; *see also* share-cropping
Thailand 6, 56, 194
transaction costs 20–1, 28, 171
tribal land and tribal chiefs 101, 108, 115, 134, 175: *see also* communal land
Tunisia: Islamic land tenure 114–5; pastoralists 116; population growth control 171–2; poverty estimates 116, 172, 185, 190; quasi complete land reform 79, 82, 114–17; rural development 115; socialist agrarian institutions 116
Turkey 57, 185

Uganda: arable land supply 53; effects on household food security 155; privatization of communally held tribal land 155–6, 214
undernutrition 5–7, 10, 12, 26, 46, 180: *see also* malnutrition; hunger
UNDP (UN Development Programme) 11–12, 38, 187, 217
UNESCO (UN Educational, Scientific and

Cultural Organization) 59, 210, 217
UNICEF (UN International Children's
Fund) 9, 217
UNRISD (UN Research Institute for Social
Development) 203, 205, 217
United Kingdom/Great Britain/Britain 26,
100, 149, 153
United Nations 148, 186; millennium
development goals for halving poverty
and hunger by 2015 1, 186–7, 217
USA: assistance for land reform in Iran
199; capitalist large farms, aims 64–5;
and Cuba 110; and Mexico 112–13;
policy shift from state administered to
market-based land reform 31, 145–8,
201, 217; share of land inheritance in
national wealth 26; South Korea 91,
195

Venezuela: effects of land concentration
on agricultural growth 60; land
concentration 55; rural poverty and
landlessness estimates 6, 184
Vietnam: food production, total and per

person 49; growth of total economy
and per person in agriculture 13; state
farms 68

wage rates in agriculture 21, 46, 75, 121,
139, 160–2
West Bengal 45, 129
WHO (World Health Organization) 196,
209
women (rural) 50, 58, 84; access to land
58–9, 175, 195, 214
World Bank classification of developing
countries 189; policy shift from
supporting state-administered to
market-based land reform 10, 148–9,
156, 163–4, 201; work on poverty 1, 4,
14, 115, 136, 139–40, 160

Yemen: among poor developing countries
11; poverty estimate 185; socialist
agrarian structure in the South 193

Zaire 183
Zambia 183

Name index

Abdalla, A. 156
Adams, M. 201
Adams, R. 125, 128, 173, 198
Adelman, I. 94
Akerlof, G. 191
Al-Dujaili, A. 196
Alamgir, M. 45
Alderman, H. 190
Alesina, A. 193
Alex, J. 139
Ali, H. 100
Ali, M.S. 174
Alwa, A. 105, 196
Aquino, C. 138
Arcand, J. 33
Assembajjwe, C. 155, 175
Atkinson, A. 38, 42
Aziz, S. 87

Backman, K. 76
Bakir, M. 109
Bandini, M. 124
Baranzini, M. 42
Bardhan, P. 78, 129, 169, 194, 199
Barraclough, S. 71, 78, 111, 193
Berry, A. 57, 73, 194
Bhalla, S. 73, 75, 194
Binswanger, H. 149, 156
Black, R. 187
Booth, A. 62
Bromley, A. 39, 42

Carter, M. 42, 163, 167, 199
Cassen, R. 14
Castro, F. 67, 145
Chayanov, A. 24
Chebil, M. 114-5

Chen, S. 158
Chenery, H. 192
Christensen, R. 78
Clark, C. 41, 192
Clark, R. 189
Cline, W. 57, 71, 73, 78, 194
Coase, R. 19, 191
Collarte, J. 193
Collier, P. 15, 72, 75, 192
Commander, S. 73–4, 127
Commons, J. 17
Cornia, G. 78, 194
Cox, T. 24
Currie, J. 62

Dandekar, V. 196–7, 199
Dasgupta, P. 62
Datt, G. 160
Deaton, A. 14
Deere, C. 67, 111
Demsetz, H. 21
Deininger, D. 32, 149, 151–2, 154, 156,
    164, 167, 171, 193
De Waal, A. 45–6
Dollar, D. 15
Dong-Wang, S. 185
Dorner, P. 78, 190
Dowson, E. 100
Duran, M. 197

El-Ghonemy, M. R. 2–7, 27, 46, 48, 62,
    65, 73–5, 83, 117, 119, 123–4, 128,
    136, 142, 151, 160–1, 169, 172–5,
    191–3, 201–2
El-Leithy, H. 173, 201
El-Zoobi, A. 141
Engle, R. 191

Ennaji, M. 137

Fields, G. 192
Forni, N. 141
Forsyth, M. 31
French, J. 194
Friedman, M. 30, 191

Gabby, R. 101
Galbraith, J. 17
Gayoso, A. 110
Georgesco-Rogan, A. 62
Ghai, D. 78, 111
Ghimire, K. 181
Ghose, A. 170
Gittinger, P. 199
Gonzalez, J. 151–2
Granger, C. 191
Gray, A. 51
Griffin, K. 69, 90, 136, 169–70, 193
Guardian, E. 200

Haider, S. 100
Hamed, H. 191
Hansen, B. 124
Harris-White, B. xv, 201
Haslett, D. 191
Hasseeb, K. 107
Hayami, Y. 72, 74, 194
Hayek, F. von 18, 30
Henry, M. 72, 74, 78, 194
Hopkins, N. 198
Hymer, S. 78

Issawi, C. 114

de Janvry, A. 154–6, 167, 189, 194
Jefferson. T. 145
Johnston, B. 192
Jose, A. 131, 199
Joshi, P. 199

Keidel, A. 91–4, 99
Kennedy, J.F. 146, 148, 201
Keynes, J.M. 192
Kuznets, S. 59

Lal, D. 72, 192
Lardy, N. 87, 195
Lazarev, G. 134, 200
Lee, E. 88, 97, 195
Lee, H. 91–2

Lenin V. 23–4
Lewis, A. 150–1
Lipton, M. 28, 191–3
Locke, J. 24
Longhurst, R. 40

MacLeod, H. 17
Mangahas, M. 139, 200–1
Mao Tse-Tung 67, 85, 157
Marshall, A. 17, 19
Marx, K. 23–4
Marzouk, G. 23–4, 124
Mellor, J. 192
Melville, B. 62
Modigliani, F. 26
Morales, E. 177, 188–9
Mosley, P. 156, 175
Muldavin, J. 158
Murray, J. 8
Myers, R. 84
Myrdal, G. 63, 191

Naschold, F. 192
Nasser, G.A. 120, 127
Nickolsky, S. 166
North, D. 191
Nouvele, J. 135
Nozick, R. 24

Otsuka, K. 155

Palmer, J. 181
Parsons, K. 27, 199, 201
Parthasarathy, G. 199
Pascon, P. 134, 137
Perkins, D. 87, 114, 157
Peters, G. xiv
Place, F. 155
Platt, K. 199
Platteau, J. 175
Putzel, J. 165, 202

Qassim, A. 177, 196
Quesney F. 51

Radwan, S. 115, 198, 201
Ram, H. 199
Ramachandran, V. xiii, 167
Ravallion, M. 45, 61
Rawls, J. 37, 177
Ricardo, D. 48
Ritter, A. 30
Roemer, J. 42, 177

Rodrik, D. 193
Roth, M. 154
Russell, B. 33

Sadat, A. 127
Saith, A. 169
Sandoval, R. 163
Sen, A. 34, 38, 44, 62, 129, 177, 192, 202
Simon, H. 191
Singer, H. xiv, 191
Smith, A. 17–8, 63
Sobhan, R. 139
Solow, R. 51
Spence, R. 191
Stigler, G. 191
Stiglitz, J. 42, 145, 191
Streeten, P. 192–3
Swearingen, W. 135
Szelenyi, I. 158, 167

Taniguchi, K. 33, 43
Taussig, M. 194
Thiesenhuesen, W. 73
Towney, H. 195
Tse-Tung, M. *see* Mao Tse-Tung

Umali. D. 203
Usher, D. 191

Visaria, P. 192
Vogelgesang, F. 152
Voelkner, H. 194

Wade, R. 43, 167, 191
Waldron, J. 43
Walker, K. 88
Wang, D. 98
Warriner, D. 27, 43, 72, 100, 198–9
White, G. 88
White, H. 187
Wolford, W. 203

Yang-Boo, C. 185
Yeltsin, B. 165–6
Young, A. 28
Yunus, M. 181

Zapata, E. 16, 177, 197
Zegara, E. 164
Zimmerman, F. 154

Routledge Studies in Development Economics

Recent and Forthcoming titles in the series include:

The Politics of Aid Selectivity
Good Governance Criteria in World Bank, U.S. and Dutch Development
Assistance
Wil Hout, Institute of Social Studies in The Hague, the Netherlands
July 2007: 234x156: 176pp
Hb: 978-0-415-37860-4: **£65.00**

Labour Standards, Development and Trade
Göte Hansson, Lund University, Sweden
July 2007: 234x156: 256pp
Hb: 978-0-415-180800-1: **£65.00**

European Union Trade Politics and Development
'Everything but Arms' unravelled
Edited by **Gerrit Faber,** Utrecht University, the Netherlands
and **Jan Orbie,** Ghent University, Belgium
March 2007: 234x156: 256pp
Hb: 978-0-415-42627-5: **£65.00**

Economic Development, Education and Transnational Corporations
Mark Hanson, University of California, Riverside, USA
July 2007: 234x156: 160pp
Hb: 978-0-415-77116-0: **£65.00**

Membership Based Organizations of the Poor
Martha Chen, Harvard University, USA, **Renana Jhabvala,** National
Coordinator of SEWA and the Chair of SEWA Bank, Ahmedabad, India, **Ravi
Kanbur,** Cornell University, USA and **Carol Richards,** Harvard University, USA
April 2007: 234x156: 320pp
Hb: 978-0-415-77073-6: **£75.00**

The Asymmetries of Globalization
Edited by **Pan Yotopoulos,** University of Florence, Itay and Professor Emeritus,
Stanford University, USA and **Donato Romano,** University of Florence, Itay
January 2007: 234x156: 208pp
Hb: 978-0-415-42048-8: **£65.00**

For further information about any of these titles please contact:
Victoria.Lincoln@tandf.co.uk or visit: www.routledge.com/economics